Biljana Arsic, Predrag Novak, Jill Barber, Maria Grazia Rimoli, Goran Kragol,
Federica Sodano
Macrolides

Also of interest

Pharmaceutical Chemistry
Volume 1: Drug Design and Action
Campos Rosa, Camacho Quesada, 2017
ISBN 978-3-11-052836-7, e-ISBN 978-3-11-052848-0

Pharmaceutical Chemistry
Volume 2: Drugs and Their Biological Targets
Campos Rosa, Camacho Quesada, 2017
ISBN 978-3-11-052851-0, e-ISBN 978-3-11-052852-7

Chemical Drug Design
Kumar Gupta, Kumar (Eds.), 2016
ISBN 978-3-11-037449-0, e-ISBN 978-3-11-036882-6

Iron-Sulfur Clusters in Chemistry and Biology
Volume 2: Characterization, Properties and Applications
Rouault, 2017
ISBN 978-3-11-047850-1, e-ISBN 978-3-11-048043-6

Iron-Sulfur Clusters in Chemistry and Biology
Volume 2: Characterization, Properties and Applications
Rouault, 2017
ISBN 978-3-11-047939-3, e-ISBN 978-3-11-047985-0

Biljana Arsic, Predrag Novak, Jill Barber,
Maria Grazia Rimoli, Goran Kragol,
Federica Sodano

Macrolides

—

Properties, Synthesis and Applications

DE GRUYTER

Authors

Dr. Biljana Arsic,
University of Nis,
Faculty of Sciences and Mathematics,
Department of Mathematics,
Visegradska 33,
18000 Nis, Serbia
ba432@ymail.com

Dr. Predrag Novak,
Department of Chemistry,
Faculty of Science,
University of Zagreb,
Horvatovac 102a,
10000 Zagreb, Croatia
pnovak@chem.pmf.hr

Dr. Jill Barber
Division of Pharmacy and Optometry
School of Health Sciences
University of Manchester
Oxford Road
Manchester, M13 9PT, UK
Jill.Barber@manchester.ac.uk

Dr. Goran Kragol,
Fidelta d.o.o.,
Prilaz b. Filipovica 29,
10000 Zagreb, Croatia
goran.kragol@glpg.com

Dr. Maria Grazia Rimoli,
Department of Pharmacy,
University of Naples "Federico II",
Naples, Italy
rimoli@unina.it

Federica Sodano,
Department of Drug Science and Technology,
University of Turin,
Turin, Italy
federica.sodano@unito.it

ISBN 978-3-11-051502-2
e-ISBN (E-BOOK) 978-3-11-051575-6
e-ISBN (EPUB) 978-3-11-051504-6
Set-ISBN 978-3-11-051576-3

Library of Congress Cataloging-in-Publication Data
A CIP catalog record for this book has been applied for at the Library of Congress.

Bibliographic information published by the Deutsche Nationalbibliothek
The Deutsche Nationalbibliothek lists this publication in the Deutsche Nationalbibliografie; detailed bibliographic data are available on the Internet at http://dnb.dnb.de.

© 2018 Walter de Gruyter GmbH, Berlin/Boston
Cover image: Foto Lik, Nis, Republic of Serbia
Typesetting: Compuscript Ltd., Shannon
Printing and binding: CPI books GmbH, Leck
∞ Printed on acid-free paper
Printed in Germany

www.degruyter.com

Preface

The first idea was to publish a short discussion on the macrolide antibiotics, but the publisher expressed an interest in a book on macrolides. Indeed, my search of the available books on the subject produced few results, some dating back to the first years of this century, and none covering all popular aspects related to macrolides. Therefore, Dr Jill Barber and I decided to ask Dr Predrag Novak to join us in this interesting journey through the world of macrolide antibiotics. He accepted, and the number of chapters increased due to his knowledge and expertise, particularly in the area of conformational analysis of macrolides and their interactions with biological targets. Dr Goran Kragol gave his word that he would write an interesting chapter on the synthesis of macrolide antibiotics with a lot of schemes and the latest approaches and strategies in this master branch of the organic chemistry. All true lovers of organic synthesis and organic chemistry will certainly like that chapter. At the end, as a strawberry on the top of the cake, Dr Maria Grazia Rimoli and Federica Sodano provided strategies for the synthesis of conjugates of macrolides with nucleobases or nucleosides. The book, therefore, gives insights on the mechanisms of action of different types of macrolides and *de facto* provides ideas on the macrolides that can be synthesized with similar characteristics as the existing macrolides or with improved properties. It also gives synthetic strategies for the synthesis of macrolide antibiotics or their conjugates with nucleobases or nucleosides.

It was indeed a huge responsibility to take on an endeavour such as writing the book in an interesting way and covering all popular topics related to macrolide antibiotics. However, from my first contact with the macrolide antibiotics I fell in love with them, so this was a quite interesting and not difficult journey for me. I hope that both students and experts will enjoy the reading journey through the book as I enjoyed taking part in its creation.

Sincerely,
Dr Biljana Arsic

https://doi.org/10.1515/9783110515756-202

Contents

Preface —— v

1	**The macrolide antibiotics and their semi-synthetic derivatives** —— **1**	
1.1	What happens to macrolides in the stomach? The effect of acid on the erythromycins and their derivatives —— **3**	
1.2	Crystal structures of macrolide antibiotics bound to bacterial ribosomes —— **9**	
1.3	Mode of action of macrolide antibiotics —— **10**	
1.3.1	Binding of azithromycin —— **13**	
1.4	Investigation of kinetics of macrolide interaction with ribosomes and possible models for different forms in binding to bacterial ribosomes —— **15**	
1.5	Solution state structures of free and bound macrolide antibiotics —— **17**	
	References —— **24**	

2	**The semisynthetic routes towards better macrolide antibiotics** —— **31**	
2.1	Introduction —— **31**	
2.2	Erythromycin —— **31**	
2.3	Second generation of macrolide antibiotics —— **34**	
2.3.1	Roxithromycin —— **34**	
2.3.2	Clarithromycin —— **35**	
2.3.3	Azithromycin —— **38**	
2.3.4	Clarithromycin/azithromycin hybrids —— **42**	
2.4	The third generation of macrolide antibiotics —— **44**	
2.4.1	Telithromycin (HMR3647) —— **47**	
2.4.2	Cethromycin (ABT773) —— **49**	
2.4.3	Solithromycin (CEM101) —— **53**	
2.4.4	Other ketolides: TE-802, modithromycin (EDP-420) —— **53**	
2.5	Antibacterial macrolide hybrids – next generations of macrolide antibiotics —— **56**	
2.5.1	Macrolide conjugates —— **57**	
2.5.2	Macrolide merged hybrids —— **57**	
2.5.3	Macrolactone ring reconstruction —— **59**	
2.6	Conclusion —— **60**	
	References —— **61**	

3	**Interactions of macrolides with their biological targets** —— **63**	
	References —— **73**	

4   **Hybrids of macrolides and nucleobases or nucleosides: synthetic strategies and biological results —— 79**
4.1   Macrolide hybrid compounds: an excellent approach —— 79
4.2   Macrolide antibiotic resistance —— 80
4.3   How and why to conjugate macrolides to nucleobases or nucleosides: introduction —— 80
4.3.1   Nucleobases or nucleosides conjugates with erythromycin and azithromycin —— 81
4.3.1.1   Introduction —— 81
4.3.1.2   Synthesis —— 81
4.3.1.3   Biological results —— 83
4.3.2   Clarithromycin-adenine and related conjugates —— 83
4.3.2.1   Introduction —— 83
4.3.2.2   Synthesis —— 83
4.3.2.3   Biological results —— 87
4.4   Novel spiramycin-like conjugates: synthesis and antibacterial and anticancer evaluations —— 88
4.4.1   Introduction —— 88
4.4.2   Synthesis —— 90
4.4.3   Antibacterial studies —— 91
4.4.4   Anticancer studies —— 92
               References —— 94

Index —— 101

Biljana Arsic[1], Jill Barber[2], Predrag Novak[3]

1 The macrolide antibiotics and their semi-synthetic derivatives

Erythromycin A (**1**) was first described in 1952, at the height of the golden age of antibiotic discovery [1]. Its structure was determined by chemical degradation, a venture that seems heroic to twenty-first century eyes [2]. The first crystal structure confirmed the results of these chemical tests [3] but showed only relative stereochemistry and, interestingly, the absolute stereochemistry was guessed incorrectly, and the wrong enantiomer of erythromycin recorded. A potent antibacterial agent, it is a product of the soil microorganism *Saccharopolyspora erythraea*. *Saccharopolyspora erythraea* produces several other erythromycins [4–9], precursors of erythromycin A and by-products of its biosynthesis (**2–7**). All are characterized by a 14-membered polyketide ring decorated with two sugars. The presence of 18 chiral centers in erythromycin A makes it a poor candidate for structure-activity relationships, and until very recently, analogues were limited to those that are found naturally and those that can be made from erythromycin in a few steps. Very recently and excitingly, a synthetic methodology has been developed that may lead to high yields of novel macrolides [10].

The British National Formulary currently recognizes just four macrolides for the treatment of bacterial infections: erythromycin, clarithromycin (**8**), azithromycin (**9**) and the ketolide telithromycin (**10**). In addition, spiramycin (**11**), which is an effective antibacterial, is used for the treatment of toxoplasmosis. Roxithromycin (**12**) is used as an antibacterial drug in a number of countries, including Australia, New Zealand and Israel. The term macrolide can be applied to other drugs, such as the immune-suppressant drugs tacrolimus and sirolimus and the polyene anti-fungals nystatin and amphotericin B, but these are beyond the scope of this chapter.

Clarithromycin, azithromycin, telithromycin and roxithromycin are all semi-synthetic derivatives of erythromycin A [11–14]. At first sight they appear to be structurally similar. Certainly, their primary mode of action is similar: they all bind to the bacterial 50S ribosomal subunit and inhibit protein synthesis, and yet small structural changes have important consequences for the actions of these drugs. For example, the semi-synthetic macrolide clarithromycin differs from erythromycin A by a single substitution of a hydroxyl group by a methoxy group, yet clarithromycin is acid-stable, whereas erythromycin is highly labile, and clarithromycin is rigid, confined almost entirely to a single conformer, whereas

1 Department of Mathematics, Faculty of Sciences and Mathematics, University of Niš, Niš, Republic of Serbia
2 Division of Pharmacy and Optometry, University of Manchester, Oxford Road, M13 9PT, Manchester, United Kingdom
3 Department of Chemistry, Faculty of Science, University of Zagreb, Zagreb, Croatia

https://doi.org/10.1515/9783110515756-001

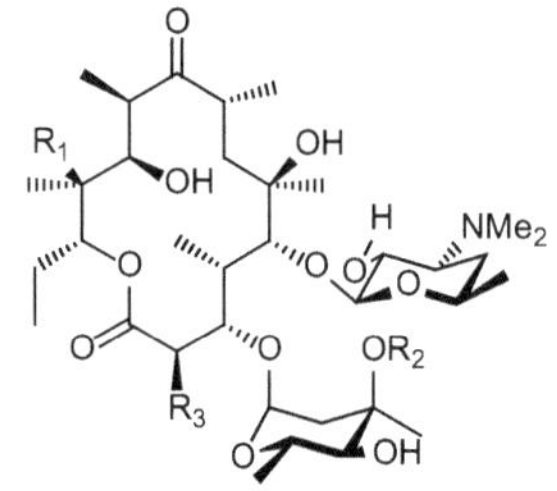

1 R₁=OH; R₂=CH₃; R₃=CH₃ erythromycin A
2 R₁=H; R₂=CH₃; R₃=CH₃ erythromycin B
3 R₁=OH; R₂=H; R₃=CH₃ erythromycin C
4 R₁=H; R₂=H; R₃=CH₃ erythromycin D
6 R₁=OH; R₂=CH₃; R₃=CH₂OH erythromycin F
7 R₁=H; R₂=CH₃; R₃=CH₂OH erythromycin G

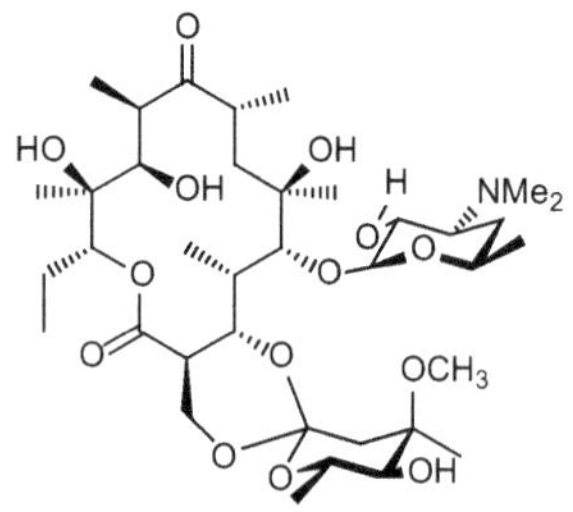

5 erythromycin E

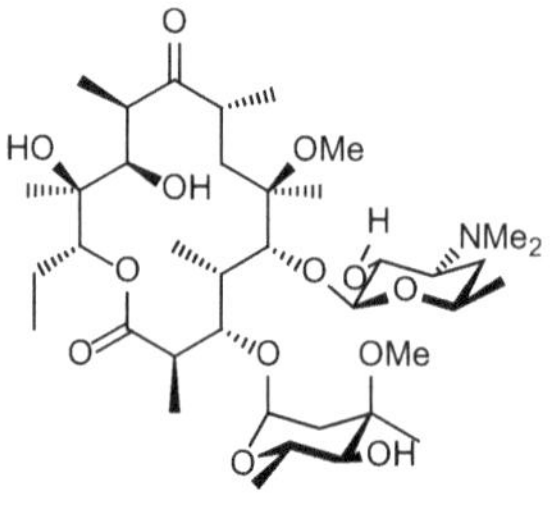

8

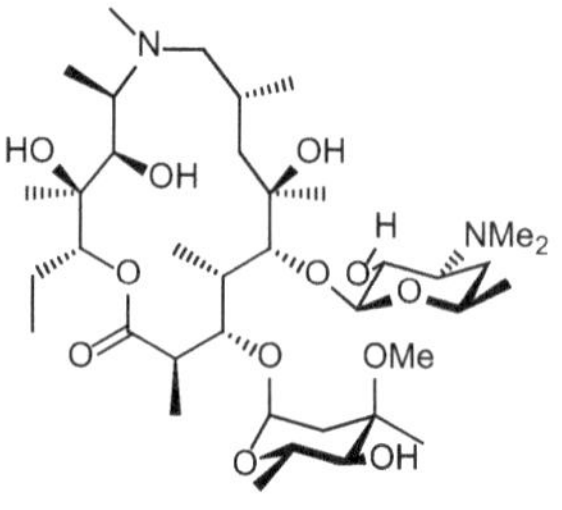

9

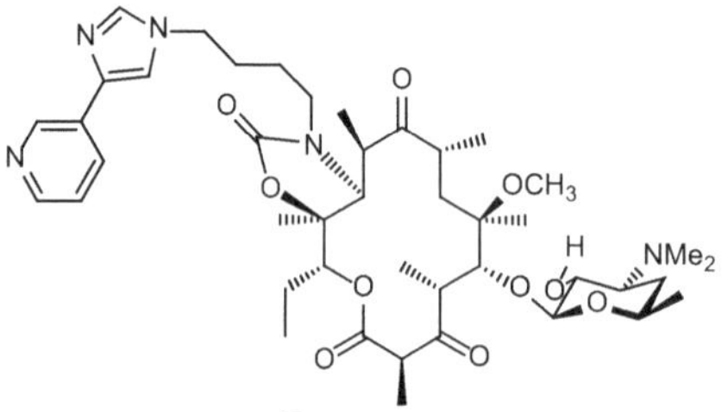

10

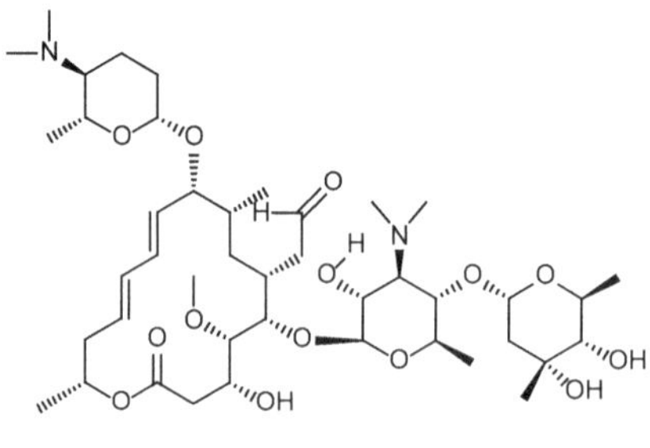

11

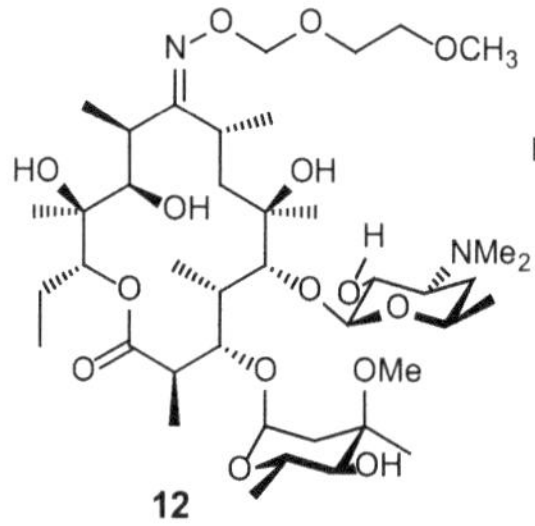

12

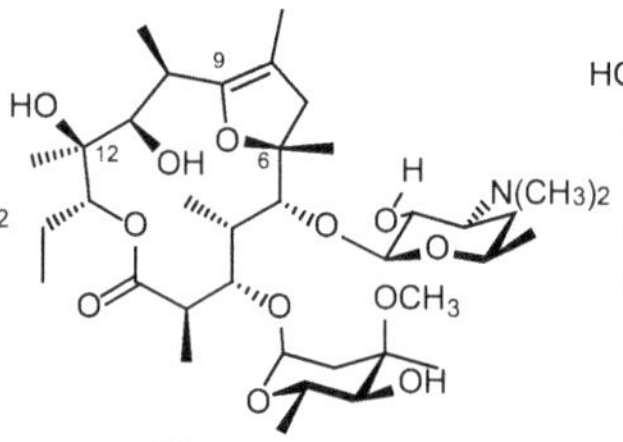

13

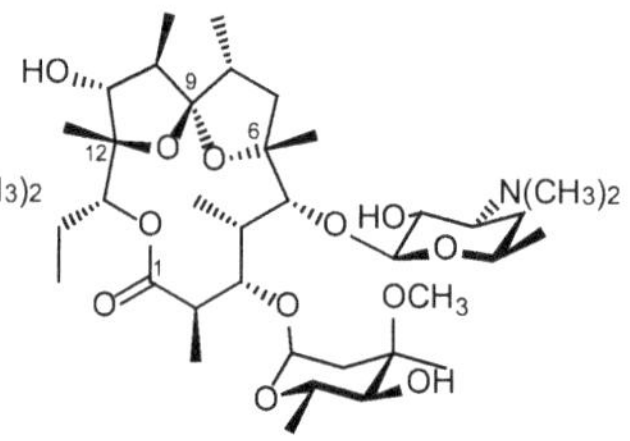

14

Scheme 1.1: Macrolide antibiotics: **1** erythromycin A, **2** erythromycin B, **3** erythromycin C, **4** erythromycin D, **5** erythromycin E, **6** erythromycin F, **7** erythromycin G, **8** clarithromycin, **9** azithromycin, **10** telithromycin, **11** spiramycin, **12** roxithromycin, **13** erythromycin enol ether, **14** anhydroerythromycin A.

erythromycin A has several minima detectable by NMR and molecular modeling. In the body, clarithromycin (**8**) and azithromycin (**9**) have long half-lives and favorable pharmacokinetics, allowing short courses (3–5 days) and daily or twice daily doses, compared with the 7-day, four doses per day standard course of erythromycin A.

In this chapter we discuss the chemistry of the macrolide antibiotics, addressing particularly the features that give them their distinctive characteristics.

1.1 What happens to macrolides in the stomach? The effect of acid on the erythromycins and their derivatives

When exposed to acid, the Achilles heel of erythromycin A proves to be the combination of the 6-OH, 9-ketone and 12-OH. Both the 12-OH and the 6-OH are capable of attacking the ketone to give 5-membered rings. Two compounds, erythromycin enol ether (**13**) and anhydroerythromycin (**14**), were isolated from the acid treatment of erythromycin A and extensive kinetic measurements were made to understand their relationship to erythromycin A and to one another. Two mechanisms were proposed as shown in Fig. 1.1 [15, 16].

A: **1** ——→ **13** ——→ **14**

B: **13** ⇌ **1** ——→ **14**

Fig. 1.1: Proposed mechanisms for the conversion of erythromycin A (**1**) to anhydroerythromycin A (**14**). A: Atkins et al. [15]; B: Cachet et al. [16].

Both mechanisms were based on kinetic data obtained using HPLC-based experiments, and were carried out before the structures of **13** and **14** had been solved unambiguously. Nuclear magnetic resonance (NMR) spectroscopy data proved important in confirming that the mechanism proposed by Cachet et al. [16] was correct. Firstly, the aqueous solution structure of erythromycin A was determined [17]. Surprisingly, this was an equilibrium mixture of the 9-ketone (**1**), as seen in the crystal structure and the 12,9-hemiacetal caused by cyclization of the 12-OH group with the ketone (**15**) (Fig. 1.2). A small amount of the 6–12 hemiacetal (**16**) could also be detected in DMSO solution. Next, the structure of erythromycin enol ether (**13**) was confirmed [18]. It was particularly important to establish that the 5-membered ring was formed by cyclization of the 6-OH (rather than the 12-OH) group with the 9-ketone. Anhydroerythromycin A was also described by NMR spectroscopy [19]. Until 2006, however, the additional chiral center (at C-9) formed by cyclization of erythromycin A to give anhydroerythromycin A had gone almost unnoticed, and the structure was generally drawn with 9S stereochemistry. X-ray crystallography and molecular modeling of anhydroerythromycin A acetate [20] revealed that the correct stereochemistry was in fact 9R, as shown in structure **14**.

The similarity of structures **14** and **15** already points towards a mechanism in which the enol ether (**13**) is formed from **16** and anhydroerythromycin A (**14**) from **15**, broadly in agreement with Cachet et al. [16]. The use of NMR, in particular the FIDDLE algorithm, allowed us to obtain very detailed kinetic measurements for erythromycin degradation [21]. The FIDDLE algorithm [22] compares the experimental time-domain signal of a reference (such as tetramethylsilane) with that predicted by theory, multiplying the raw experimental data by the complex ratio of the two signals to produce a corrected free induction decay (FID); this results in excellent line shapes in the NMR spectrum from which kinetic parameters can be deduced.

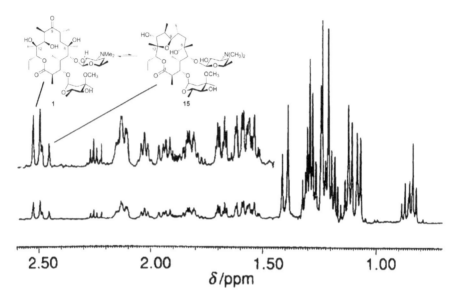

Fig. 1.2: Part of the 500 MHz ¹H NMR spectrum of erythromycin A in D₂O, showing the presence of two interconverting isomers.

When the intensities of representative signals from **1, 13** and **14** were plotted (Fig. 1.3) it could be seen, because of the high quality of the data, that the Cachet et al. model did not perfectly describe the concentrations of these compounds at long times. Incubation of anhydroerythromycin A in acid conditions gave rise to small amounts of erythromycin A and erythromycin enol ether, demonstrating that **14**, as well as **13**, equilibrates with erythromycin A [23]. The loss of the cladinose sugar could also be detected. These results were modelled mathematically and, taken together with those of Volmer and Hui [24], give the complex scheme shown in Fig. 1.4, representing the acid-catalysed degradation of erythromycin A.

These studies illustrate the possibilities of carrying out kinetic measurements of drug degradation by NMR spectroscopy; however, clinically, the most important finding was the simplest – that erythromycin A has a very short half-life (<20 minutes) in acidic conditions. The pediatric pro-drug erythromycin A 2′-ethyl succinate (**18**) degrades with almost identical kinetics [23] and escapes unacceptable degradation in the stomach by virtue of being sparingly soluble and being formulated to leave the stomach quickly.

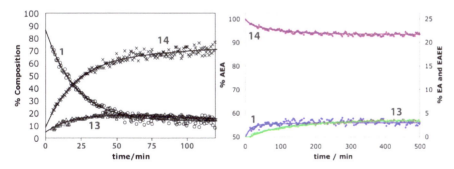

18

The semi-synthetic antibacterial drug clarithromycin (**8**) [11] is structurally closely similar to erythromycin, yet its properties, both *in vitro* and *in vivo*, are remarkably different. In acid, the drug degrades by loss of the cladinose sugar [25, 26] to give **19**, with a half-life of 310 minutes at 37°C [27]. On a physiological timescale, nothing else of interest takes place, except that NMR spectroscopy indicates the fleeting presence of a number of minor components. One of these may be the 12,9-hemiacetal (**20**), which might be expected by analogy with an aqueous solution of erythromycin A. This compound was not detectable by NMR in neutral solution [28]. When the acid-catalysed degradation was allowed to continue, two further compounds were detected, in equilibrium with one another. Structures **21** and **22** had been proposed by Nakagawa et al. [26] as products of the acid treatment of **20**, and the NMR data [27] were consistent with this suggestion (Fig. 1.5).

Fig. 1.3: Time courses of the acid-catalysed treatment of erythromycin A and anhydroerythromycin A.

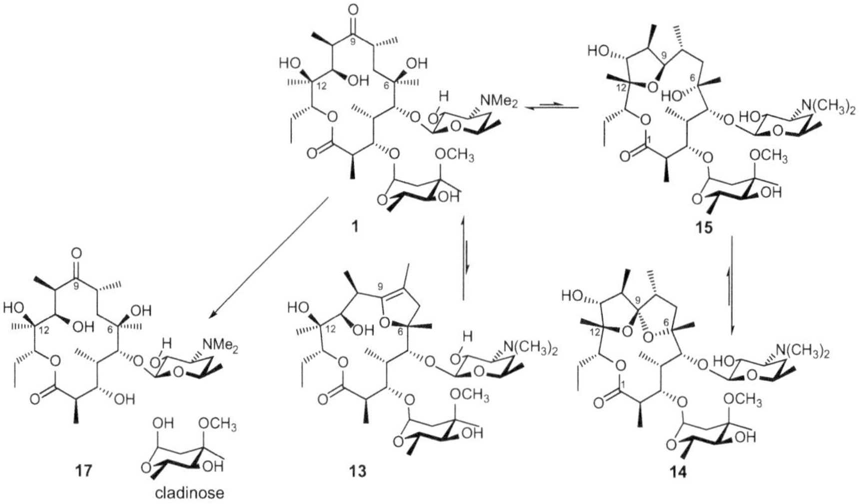

Fig. 1.4: Pathway for the acid-catalysed degradation of erythromycin A, after Hassanzadeh et al. [23].

The degradation in acid of azithromycin was studied by Fiese and Steffen in 1990. They concluded that, like clarithromycin, azithromycin degrades by loss of the cladinose sugar. The conditions were different from those used to study clarithromycin so a direct comparison of the rates of reaction is not possible. It is possible to say, however, that azithromycin is essentially stable to stomach acid. Roxithromycin (**12**) similarly has C9 unavailable for cyclization reactions, and presumably degrades slowly *via* loss of the cladinose sugar, although there is no literature directly supporting this.

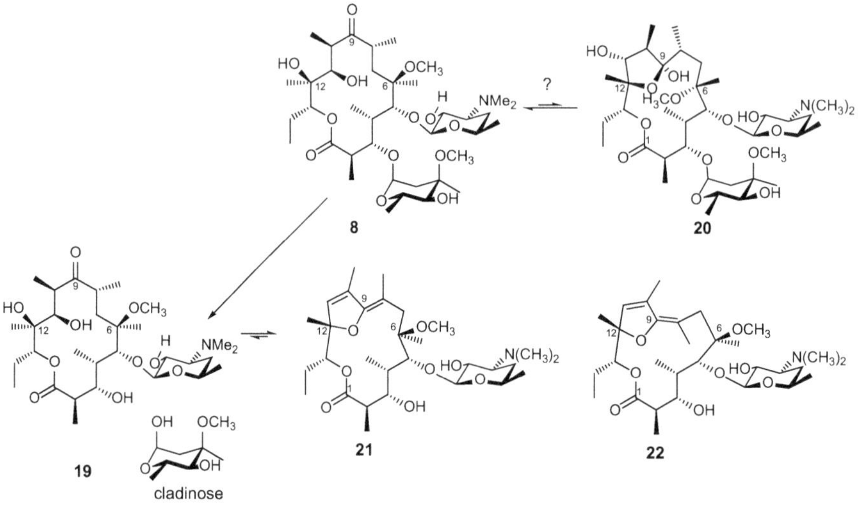

Fig. 1.5: Proposed pathway for the acid-catalysed degradation of clarithromycin (**8**).

The data for degradation of telithromycin are sparse. Traunmüller et al. [29] carried out high performance liquid chromatography of extracts from human plasma and developed a method of quantifying telithromycin in plasma. Their conclusions about its stability were, however, qualitative: "Telithromycin was found to be unstable in aqueous solution at ambient temperature". Hydrolysis of the carbamate side chain was postulated as a degradative mechanism. Telithromycin lacks a cladinose sugar, and the normal method of degradation *via* loss of this sugar is therefore impossible. The literature on telithromycin degradation is motivated mainly by pharmaceutical concerns rather than by chemistry and the identification of telithromycin degradation products has not been a high priority.

The remaining macrolide that has been studied in detail is erythromycin B [27]. This compound (**2**) is a side-product of erythromycin A production and lacks the 12-OH group. It therefore cannot cyclize in a 12-9 direction. Azithromycin and roxithromycin both lack the 9-ketone, and clarithromycin has a modified 6-OH group. Unsurprisingly, the ultimate degradation pathway for erythromycin B was found (by diffusion ordered spectroscopy (DOSY)) to be the descladinose compound, **23**. Much more surprising was the mechanism by which this compound was obtained. Fig. 1.6 shows the time-course of the degradation of erythromycin B at apparent pH 2.5 and 37°C in deuteriated buffer, as monitored by NMR spectroscopy, again using the FIDDLE algorithm [22, 27]. The unusual time-course results from the equilibrium of erythromycin B with its 6,9-enol ether (**24**). Dehydration of erythromycin B involves breaking the C8-H bond. When this re-forms, the deuteriated solvent supplies a deuterium rather than a protium and 8-*deuterio*- erythromycin B (**2d**) then dehydrates about 5 times more slowly than the parent erythromycin B (**2**).

There are several interesting features of this reaction. Firstly, the half-life of the loss of the cladinose sugar to give **23** is around 280 minutes, comparable with clarithromycin, not with erythromycin A (which has a half-life of below 10 minutes in these conditions). Erythromycin B has been reported to have comparable [30] or slightly lower [31] antibacterial activity compared with erythromycin A, and this may have been the incentive for the optimization (by classical means of strain selection) of *S. erythraea* to produce erythromycin A at the expense of the other erythromycins. With hindsight, we may ask whether erythromycin B, with its much improved acid stability, would in fact have been the better compound.

The second point of particular interest is that, although erythromycins A and B both form 6,9-enol ethers in acidic solution, the position of equilibrium is very different. For erythromycin A, the equilibrium lies on the side of the enol ether (the inactive compound). By contrast, in the case of erythromycin B, the equilibrium is on the side of the active drug. Attempts have been made to exploit this equilibrium.

Erythromycin B enol ether ethyl succinate (**25**) is an almost insoluble derivative of erythromycin B, stable at neutral pH [32]. In acid, the enol ether ring opens to yield erythromycin B ethyl succinate (**26**), which is analogous to erythromycin A ethyl succinate (pediatric erythromycin). This compound hydrolyses to erythromycin

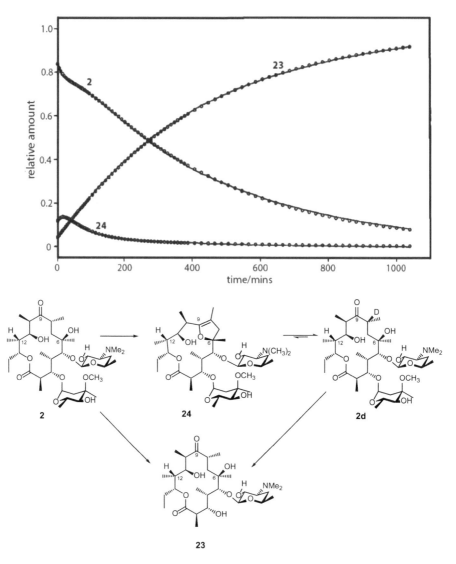

Fig. 1.6: Top – kinetics of degradation of erythromycin B at apparent pH 2.5 and 37 °C; bottom – degradation pathway showing the origin of the isotope effect in the formation of **24**.

B in neutral or basic conditions. Erythromycin A ethyl succinate (**27**) is a pro-drug for erythromycin A and hydrolyses to give the active drug in the intestine and blood-stream. Unfortunately, clinical pediatric suspensions are also susceptible to hydrolysis on storage, and erythromycin has a foul, bitter taste. Some children gag even on freshly made up erythromycin suspensions, which always contain a few percent erythromycin. The advantage of erythromycin B enol ether ethyl succinate is that it is so insoluble in neutral solution that it is stable for weeks. In acid (such as the stomach) it dissolves and converts almost instantly to erythromycin B ethyl succinate (Fig. 1.7). This relies on the equilibrium between the erythromycin and its enol ether being on the side of the ring-opened drug, and the same chemistry cannot be applied to erythromycin A.

In a second attempt to exploit this equilibrium for clinical benefit, 8-*deuterio*-erythromycin B (**2d**) has been prepared [33]. The rationale here is that erythromycin enol ethers are reported to be largely responsible for gut motilide side-effects of erythromycins [34–36], and deuteriation helps to suppress the formation of enol ethers.

Unfortunately, the pharmaceutical industry is heavily committed to erythromycin A (which is also the precursor to clarithromycin, azithromycin, roxithromycin and telithromycin), and the relatively small improvements afforded by using erythromycin B are unlikely to be economical. Clarithromycin B (**28**) has been made, but showed no advantages over the conventional clarithromycin A [37].

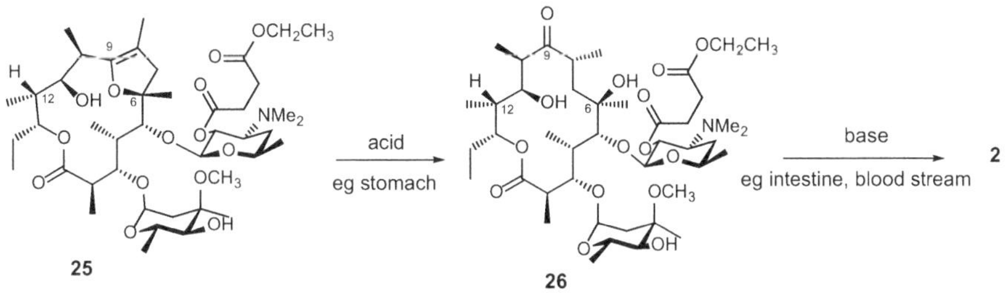

Fig. 1.7: The conversion of erythromycin B enol ether ethyl succinate (**25**) to erythromycin B ethyl succinate (**26**) and thence to erythromycin B (**2**).

1.2 Crystal structures of macrolide antibiotics bound to bacterial ribosomes

During the first decade of this century, crystal structures for several antibiotics bound to the large ribosomal subunits were published. Yonath et al. generated the structures of *Deinococcus radiodurans* large ribosomal subunit in complex with erythromycin and telithromycin (14-membered macrolides), as well as lincosamide, clindamycin, and dalfopristin and quinupristin (streptogramins of the A and B types, respectively) [38–40]. The structures of azithromycin (15-membered macrolide) and carbomycin, spiramycin and tylosin (16-membered macrolides) and virginiamycin M (A-type streptogramin) bound to the large ribosomal subunit from *Haloarcula marismortui* have also been determined by Steitz group [41, 42].

Crystal structures of antibiotics bound to ribosome presented by Tu et al. [43] and Hansen et al. [41] differ significantly from structures provided by the Yonath group

[38–40] although *H. marismortui* and *D. radiodurans* large subunits are highly conserved. The observed differences can be divided into three categories: 1) drug conformation; 2) variations in the positions adopted by drugs of the same class in a bound state; and 3) drug-ribosome interactions.

Crystal structures of numerous antibiotics alone were determined previously. The structures of these antibiotics when they are bound to *H. marismortui* large subunit are almost identical to the conformations they show as single compounds. In case of conformations with the same molecules in complexes with *D. radiodurans* large subunit, they are quite different. Examples are erythromycin where cladinose sugar is in low-energy chair conformation both in small molecule crystal structure and complex provided by Stephenson et al. [44], but in complex with large subunit provided by Schlunzen et al. [40] it is in the boat conformation (Tab. 1.1).

Tab. 1.1: Comparison of structures reported for antibiotics bound to large ribosomal subunits with the structures available in the Cambridge Structural Database [43].

Antibiotic (CSD[a] number)	*H. marismortui*	*D. radiodurans* (PDB number)
Erythromycin (NAVTAF)	0.745[b]	2.307 (1JZY)
Clindamycin 1[c] (SUPBIO)	0.312	2.100 (1JZX)
Clindamycin 2	1.017	1.868
Clindamycin 3	0.574	2.387
Telithromycin (GOPGAT)	0.260	1.799 (1P9X)
Virginamycin S (KEFWUN)	0.779	1.733 (1SM1)

[a] Cambridge Structural Database.
[b] Root mean square deviations are reported (in Å) for the optimal superposition of the structures being compared. Only atoms common to both the small molecule structures and the antibiotic complex structures reported are aligned and compared.
[c] The different conformations observed for clindamycin 2-phosphate are indicated with numbers.

1.3 Mode of action of macrolide antibiotics

The effect of macrolides is a result of sterical blocking of the lumen of the tunnel [41], particularly for larger macrolides. However, recent biochemical data show that the steric block model is incomplete for some macrolides (erythromycin). Namely, erythromycin induces the dissociation of peptidyl-tRNAs from the ribosome [45].

There is only a general knowledge on the interaction of macrolides with ribosomes, if we consider tight binding of antibiotics to bacterial ribosomes. Erythromycin class antibiotics do not block peptidyl transferase activity [46]. They are binding

to the peptidyl transferase ring and block the tunnel that channels the nascent peptides away from the peptidyl transferase center [47]. Erythromycin A, roxithromycin
and clarithromycin bind to the same site in the 50S subunit of *D. radiodurans* at the
entrance of the tunnel (Fig. 1.8).

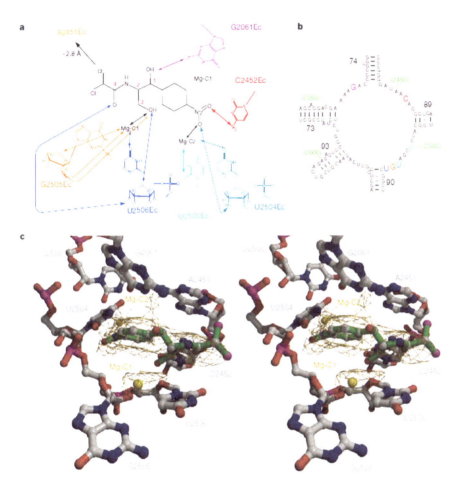

Fig. 1.8: Interaction of macrolides with the peptidyl transferase cavity. a) Chemical structure
diagram of macrolides (erythromycin A, clarithromycin and roxithromycin) showing the
interactions of the reactive groups of macrolides with the nucleotides of peptidyl transferase
cavity. b) Secondary structure of 23S rRNA with marked contacts of the medicine with RNA moiety.
c) Stereoview showing the erythromycin binding site at the peptidyl transferase cavity of
D. radiodurans in the scheme, the antibiotic is represented by the green part of ribosomal protein
with yellow, part of ribosomal protein L22 with light green. Nucleotide numbering is according to
the *E. coli* sequence [40].

The desosamine sugar and lactone ring take part in the interaction of the drug with the peptidyl transferase cavity [48]. The desosamine sugar 2'OH group forms hydrogen bonds with three positions: N6 and N1 of A2041Dr (A2058Ec) and N6 of A2042Dr (A2059Ec) [40]. Hydrogen bonding interactions of the hydroxyl group of desosamine explain resistance mechanisms – the dimethylation of the N6 group not only adds a bulky substituent causing steric hindrance for the binding, but prevents the formation of hydrogen bonds to the 2'OH group. Several hydroxyl groups have an interaction with 23S rRNA in the lactone ring: the 6-OH (hydrogen bonding to N6 of A2045Dr (A2062Ec)) [49], 11-OH and the 12-OH groups (both form hydrogen bonds with O4 of U2588Dr (U2609Ec)) [50]. The cladinose sugar does not take part in interactions with 23S rRNA.

A common pattern to the interaction and action of 14-membered macrolides to bacterial ribosomes is interactions between N6 of A2041Dr (A2058Ec), N6 of A2042Dr (A2059Ec) and the non-bridging phosphate oxygen of U2484Dr (U2505Ec). In order to prevent binding to peptidyl-tRNA, a nucleotide is situated at the G2484Dr (G2505Ec) position [51].

The macrolide binding site can be found at the entrance of the exit tunnel (Fig. 1.9).

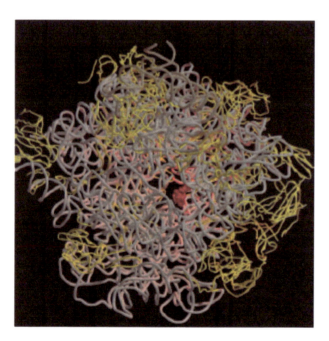

Fig. 1.9: Top view of the *D. radiodurans* 50S subunit showing erythromycin (red) bound to the entrance of the exit tunnel. Ribosomal proteins are represented by yellow color, 23S rRNA by gray and 5S rRNA by dark gray [40].

In the bound state, the drug reduces the exit tunnel diameter from approximately 18 Å to nearly 10 Å. The exit tunnel is occupied also by the hydrated Mg²⁺ ion. Neither drug nor magnesium show direct interaction with ribosomal proteins.

Azithromycin shows a difference from other macrolide antibiotics having the secondary binding site (Figs. 1.10 and 1.11).

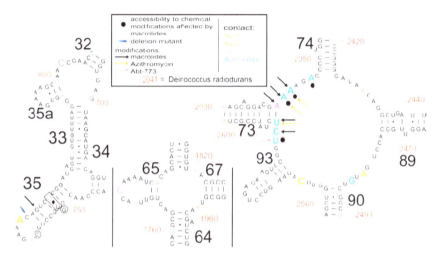

Fig. 1.10: The regions of domains II, IV and V of 23S rRNA that contribute to binding. The contact sites of ABT-773 and azithromycin are indicated by large, colored letters in the diagram [52].

1.3.1 Binding of azithromycin

There are three available co-crystal structures of azithromycin with large ribosomal subunits of three quite different bacteria (*Thermus thermophilus* [53], *H. marismortui* [41] and *D. radiodurans* [52]). In the first two, one azithromycin molecule is present in the crystal structure, and in the last, two molecules of azithromycin.

Structures of azithromycin in complexes with large ribosomal subunits of *T. thermophilus* and *H. marismortui* share a common set of hydrophobic interactions between the lactone ring of the macrolide and the hydrophobic surface of the ribosomal exit tunnel (U2611, A2058, and A2059).

The primary site of azithromycin binding to *D. radiodurans* is located in domain V of 23S rRNA with similar orientation to erythromycin A. The second molecule forms the hydrogen bond between O4 of U2588Dr with a direct contact observed with the primary binding site. It seems that the second molecule of azithromycin is in the interaction with the L4 ribosomal protein. Additional contacts with domains II and IV lead to another region of 23S rRNA, giving as a result tighter binding. Present hydrophobic interactions

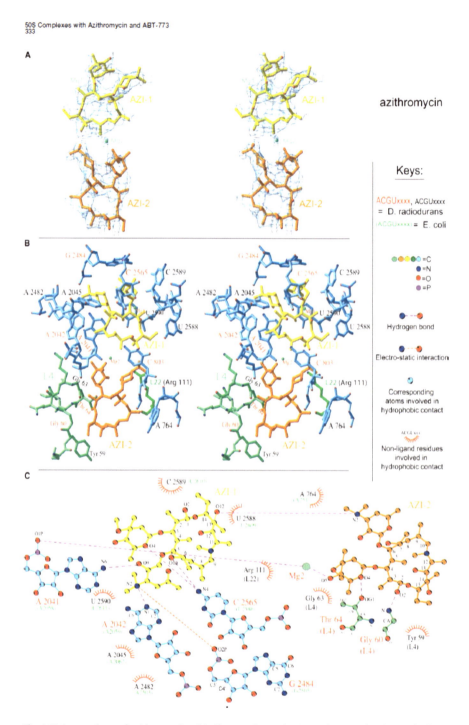

Fig. 1.11: Interactions of azithromycin with ribosomal proteins L4 and L22 and 23S rRNA [52].

of bases 2045Dr (A2062Ec) and U2588Dr (U2609Ec) have a contribution to the binding of the first azithromycin molecule. A putative Mg ion is in coordination with water molecules, and it interacts with the cladinose sugar and lactone ring. Interestingly, the nitrogen atom in the lactone ring of azithromycin does not contribute directly to the ribosome binding. The second molecule of azithromycin is making a direct contact with the first molecule of azithromycin with the formation of a hydrogen bond between the desosamine sugar and O1 in the lactone ring of the first azithromycin molecule.

1.4 Investigation of kinetics of macrolide interaction with ribosomes and possible models for different forms in binding to bacterial ribosomes

Goldman et al. [54] found that forward rate constants and binding affinity do not show any correlation with pH when the interaction of erythromycin with ribosomes (from Gram-positive and Gram-negative bacteria) was investigated, leading to the conclusion that the protonated form of this macrolide binds to ribosomes. Similar results were obtained with azithromycin. Therefore, an increase of the extracellular pH increases the antimicrobial potency of basic [55] and not neutral [56] macrolides, and consequently increases the rate of uptake into the bacterial cell [57]. It was determined that the dissociation rates for the macrolide-ribosome complexes followed a first-order rate at pH 7.2, and they were between 0.138 and 0.063 min^{-1} for erythromycin and 0.05 and 0.019 min^{-1} for azithromycin (Tab. 1.2). It is noteworthy that erythromycin dissociates faster from Gram-negative ribosomes compared to Gram-positive [54].

Pestka et al. [58] concluded that antibacterial activity depends not only on the binding of the analogue to ribosomes, but also on its ability to enter and function intact in these cells.

Tab. 1.2: Kinetics of macrolide interaction with ribosomes at pH 7.2 [54].

Ribosome source	Kinetic constants for[a]:					
	Erythromycin			Azithromycin		
	k_1	k_{-1}	K_d	k_1	k_{-1}	K_d
H. influenzae	6.0×10^7	0.106	1.8×10^{-9}	5.2×10^7	0.016	3.1×10^{-10}
E. coli	6.4×10^7	0.138	2.2×10^{-9}	6.4×10^7	0.018	2.8×10^{-10}
B. subtilis	4.9×10^7	0.063	1.3×10^{-9}	3.1×10^7	0.050	1.6×10^{-9}
S. aureus	2.5×10^7	0.024	9.6×10^{-10}	2.8×10^7	0.019	6.8×10^{-10}

[a] All reactions were performed in ribosome-binding buffer at pH 7.2, and samples were taken in triplicate for determination of the amount of macrolide bound to ribosomes by filter binding. Kinetic constants: k_1, forward rate constant (liters per mole per minute); k_{-1}, reverse rate constant (per minute); K_d, dissociation constant (molar concentration).

A

$$D^0 + R \Leftrightarrow [D^0R] + H^+ \Leftrightarrow D^+R$$

$$D^+ + R \Leftrightarrow D^+R$$

B

$$D^+ + R \Leftrightarrow [D^+R] \Leftrightarrow D^0R + H^+$$

$$D^0 + R \Leftrightarrow D^0R$$

C

$$D^0 + R \Leftrightarrow D^0R$$

$$D^+ + R \Leftrightarrow D^+R$$

D

$$\text{I] } D^0 + R \Leftrightarrow D^0R \quad D^+ + R \quad X \Rightarrow$$

$$X = \text{no binding}$$

$$\text{II] } D^+ + R \Leftrightarrow D^+R \quad\quad D^0 + R \quad X \Rightarrow$$

Fig. 1.12: Possible models for the roles of neutral and protonated forms of macrolides in binding to bacterial ribosomes. D^0, Neutral macrolide; D^+, protonated macrolide; R, ribosome; $X \Rightarrow$, no binding. Complexes in brackets represent loosely bound intermediates [54].

A postulate exists that erythromycin A and some of its derivatives bind to the ribosome of *Escherichia coli* in a two-stage process. A strong binding interaction was detected by equilibrium dialysis and related techniques [54, 59, 60], including X-ray crystallography [43]. A weak interaction has been characterized as a fast exchange process by NMR spectroscopy [61–63]. Unlike the strong interaction, the weak interaction appears to be magnesium independent [64].

The association and dissociation of macrolides can be described by the general formula $D+R \leftrightarrow DR$, where D represents a macrolide and R represents a ribosome; the forward and reverse rate constants are k_1 and k_2, respectively. As macrolides containing desosamine exist in both protonated and neutral forms, the involvement of both forms in the binding reaction must be considered. Goldman et al. [54] considered several possibilities: 1) Both forms bind similarly to ribosomes, and both $D^0+R \leftrightarrow D^0R$ and $D^++R \leftrightarrow D^+R$ reactions occur (D^0 and D^+ are the neutral and protonated forms of the macrolide, respectively), with k_1 and k_2 being similar for both D^0 and D^+ (Fig. 1.12C). 2) Both protonated and neutral forms bind to ribosomes with similar affinities and with slightly different rate constants. 3) Both protonated and neutral forms initiate binding to ribosomes, but only the protonated (Fig. 1.12A) or neutral (Fig. 1.12B) form is present in the tightly bound macrolide-ribosome complex. In case A (Fig. 1.12A), only the protonated form is considered to bind directly into the tight complex, with the neutral form evolving from a loosely bound intermediate that locks into the tightly bound form after protonation. In case B (Fig. 1.12B), only the neutral form is considered to bind directly into the tight complex, with the protonated form evolving from a loosely bound intermediate that locks into the tightly bound form after deprotonation. 4) Only the neutral or only the protonated macrolide form initiates binding and performs the final tight binding (Fig. 1.12D).

1.5 Solution state structures of free and bound macrolide antibiotics

In addition to crystal structures, steps taken in the process of drug design should also include elucidation of the solution-state structures of free and bound ligand molecules as the structural features of the complex may not be exactly the same in solution and in the solid state.

Accurate determination of free-state solution 3D structures of small-molecule ligands provides an efficient tool for medicinal chemistry to better understand binding properties [65]. Structurally, macrolides are complex molecules with many stereogenic centers that affect their biological and physico-chemical properties. This should be kept in mind when designing libraries of potentially bioactive macrolides or novel classes of antibiotics [66, 67]. Interactions with bacterial membranes and bile acids are also crucial for the overall macrolide biological profile and can be connected to macrolide transport and bacterial resistance [68, 69]. This can have a significant impact on the development of new strategies aimed at the design of compounds with enhanced biological activity.

The interactions of macrolides and bacterial ribosomes can be studied by different methods and a number of experimental approaches have been developed to characterize binding such as filtration, RNA footprinting, peptidyl transferase activity and surface plasmon resonance, for example [70–75]. Many of them need radiolabelled antibiotics as a ligand and do not provide mechanistic details of those interactions at the molecular level. Changes in fluorescence intensity and fluorescence polarization can also be used to characterize the binding of macrolides, but fluorescently labelled ligands are required [76–78].

Crystallography represents a powerful tool to characterize binding at the atomic level provided suitable crystals are obtained. Despite the fact that important structural information on macrolide-ribosome complexes were obtained, most of the structures were observed for ribosomes isolated from clinically non-relevant bacteria, except for *E. coli*, and cannot explain all the effects of macrolides on different pathogenic strains. However, solid and solution-state structures may not be the same. Hence, steps taken in the process of drug design should take into account solution-state structures of free and bound ligand molecules as well.

The 14- and 15-membered macrolide compounds are considered to be the most important in medicinal chemistry, and hence, the majority of conformational studies have been performed for these compounds (Fig. 1.13).

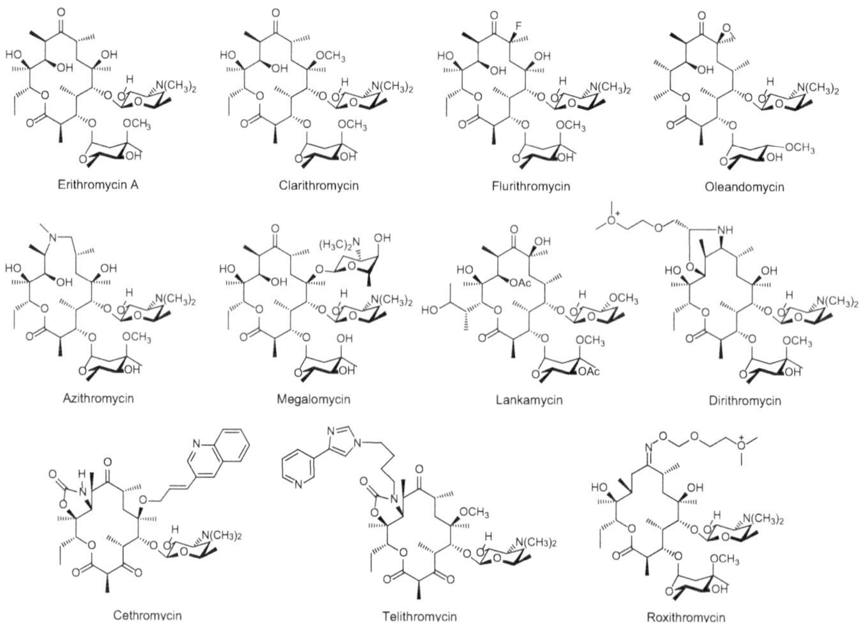

Fig. 1.13: Structures of the main representatives of 14- and 15-membered macrolides.

The complex structures of the macrolides suggest the variety of possible conformations and the ways certain structural parts can be spatially oriented that affect not only their ability to bind to ribosomes but also their physico-chemical characteristics. Hence, crystalographic data revealed similar ribosome binding modes of these antibiotics, but they may significantly differ in their pharmacokinetic properties such as dissolution, intestinal absorption and cellular permeability, for example.

Most of the macrolide-ribosome crystal structures pointed towards the folded-out macrolide conformation as the bound one [41]. In this conformation, the polar groups of the macrocyclic lactone ring are directed towards one side of the molecule, whereas the other side is mostly hydrophobic. When bound to ribosome, the polar macrolide region is in close contact with the lumen of the exit tunnel while the hydrophobic part is oriented towards the tunnel wall [41, 43].

Everett et al. [79] described that erythromycin A, the first isolated macrolide antibiotic [80], exists in two main conformational states, the folded-out and the folded-in, referring to the outward and inward folding of the macrolactone ring fragment C3–C5, as shown for azithromycin in Fig. 1.14. The folded-out conformation is based on the crystal structure of erythromycin A hydroiodide [3], whereas the folded-in is based on the crystal structure of dirithromycin [81]. In the folded-in conformation, atom H3 approaches H11 and atom H4 approaches Me-18, whereas in the folded-out conformation atom H4 approaches H11 and H5 approaches Me-18 (Fig. 1.15).

A comprehensive study of the solid-state conformation of the 14-membered macrolide antibiotics erythromycin A and B, clarithromycin and roxithromycin has been undertaken by Miroshnyk et al. [82]. It has been shown that the macrocyclic ring adopted a folded-out conformation in all of the studied macrolides, thus demonstrating its rigidity. The solid-state conformations of clarithromycin and especially roxithromycin were shown to differ from those of erythromycin A and B in the relative orientation of their monosaccharide moieties, cladinose and desosamine. The different spatial arrangements of these sugar units were underlined as factors that should be taken into account when establishing the relationships between the molecular structure and pharmacokinetic properties of the macrolide antibiotics.

Many approaches to understand the conformations of macrolides in solution have been used over the years. Most of them have involved a combination of NMR and molecular modeling methods. The NMR data used include coupling constants, in the macrolide ring and sugar units, and nuclear Overhauser effect (NOE) data. The conformational behavior of 14- and 15-membered macrolide antibiotics such as erythromycin A, roxithromycin, clarithromycin and azithromycin has been studied intensively by NMR and modeling [48, 65, 83–91]. Some studies have reported NMR constraints applied in molecular modeling calculations [48, 87, 92], and in some, molecular modeling calculations were carried out unconstrained and the obtained structures compared with the NMR data [66, 88–91]. One of the problems associated with using constraints in the calculations is that artificial average conformations that do not actually exist may be obtained [91].

The solid and solution state conformations of various azithromycin derivatives and their aglycones were determined by X-ray diffraction [93–95] and NMR spectroscopy [65, 91, 96]. Those investigations also demonstrated that these compounds exist in two major conformational families: folded-out and folded-in, as previously determined for erythromycin derivatives. The folded-out conformers have larger homonuclear $^3J_{H2,H3}$ values (~10 Hz), larger torsion angles between atoms H2 and H3 (~±180°) and exhibit a close space approach of protons H4 and H11, giving rise to NOE or ROE cross peaks (Fig. 1.15).

Much lower $^3J_{H2,H3}$ values (~2–3 Hz), lower torsion angles (~100°) and a close space approach of atoms H3 and H11, with the corresponding NOE cross peaks are characteristic of the folded-in conformers [79, 83, 86].

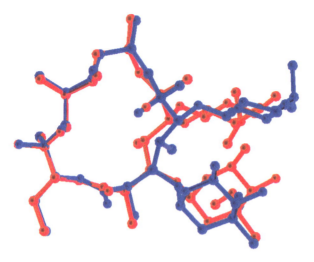

Fig. 1.14: Superposition of the folded-out and folded-in conformations of azithromycin.

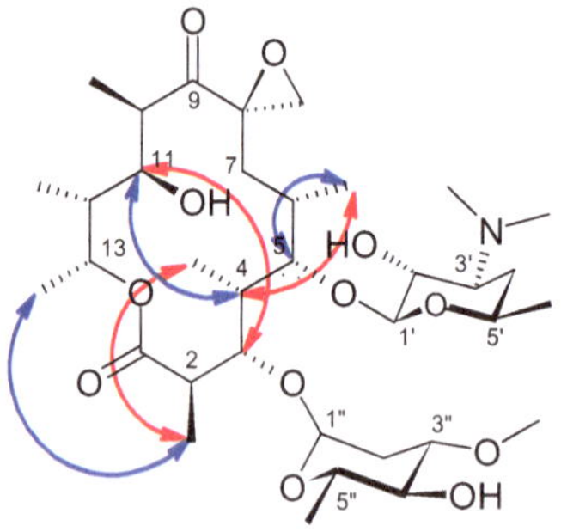

Fig. 1.15: Key NOE contacts for the folded-in (red) and folded-out (blue) conformations in oleandomycin.

Desosamine and cladinose sugars in 14- and 15-membered macrolides are oriented closer to each other in the folded-out conformation than in the folded-in conformation. In most cases they were found to adopt energetically favored chair conformations [83]. For oleandomycin and its derivatives, both conformations were detected by NMR in solution whose amount depended on the solvent polarity and temperature [89, 91]. The conformation around the glycosidic bonds, governing the relative orientation of sugars *vs.* the lactone ring, showed certain flexibility within two conformationally close families. Some conformational flexibility was also observed in the erythronolide part of the molecule. It was concluded that the X-ray structures do not adequately describe the solution-state conformations of oleandomycin and its derivatives (Fig. 1.16).

Novak et al. [88, 91, 97] have demonstrated that an approach which combines NMR and molecular modeling methods could be applicable to conformational studies of

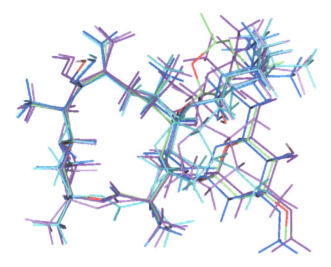

Fig. 1.16: Superposition of different conformers of oleandomycin derivatives.

free and bound macrolides and their interactions with ribosomes. Their results, and those of other groups [36, 78, 83, 86, 98, 99], have confirmed that macrolides adopt two major conformational states, folded-out and folded-in, referring to the outward and inward folding of the ring fragment 3C-5C.

In the case of macrolides, conventional nuclear Overhauser enhancement spectroscopy (NOESY) experiments are sometimes subject to errors because macrolide NOEs are close to zero at commonly used field strengths and solvents. For example, azithromycin displayed a nice NOESY spectrum at 500 MHz spectrometer in $CDCl_3$, but only a few NOEs were detected in the NOESY spectrum in D_2O. Hence, in such cases the rotating frame experiments (ROESY) can be used which give rise to equally weighted positive signals but are technically demanding and very susceptible to artefacts (Fig. 1.17).

It was concluded that the vicinal coupling constants $^3J_{H2,H3}$ and NOE cross peaks between protons H3-H11 and H4-H11 were good indicators of aglycone ring folding. Furthermore, $^3J_{CH}$ coupling constants over the glycosidic bonds provided information about the position and mobility of sugar units, with respect to the lactone ring. Longitudinal relaxation of methyl protons could be useful to probe motions of methyl groups which reflect aglycone ring folding. A conformation intermediate between the classical folded-out and folded-in, termed 3-*endo* folded-out for some 6-*O*-methyl homoerythromycin derivatives has also been reported [90]. This conformation was characterized by $^3J_{H2,H3}$ of approximately 4–5 Hz and NOE interactions between H11 and both H3 and H4.

Studies performed by Ćaleta et al. [99], which included a combined use of X-ray, NMR and molecular modeling methods, have demonstrated that macrolide aglycons of 14- and 15-membered macrolide classes, *e.g.* (*E*)-9-hydroxyimino-6-*O*-methylerythronolide A,

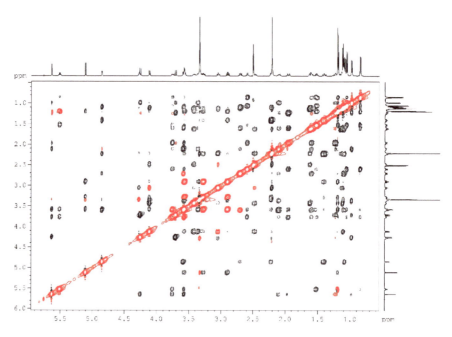

Fig. 1.17: 2D ROESY NMR spectrum of oleandomycin derivative in DMSO-d_6.

9a-aza-9-deoxo-9,9-dihydro-9a, 11-O-dimethyl-9a-homoerythronolide A and 9a-aza-9-deoxo-9,9-dihydro-9a-homoerythronolide A also predominantly adopted the folded-out conformation in solution and in the solid state. They concluded that the macrolactone ring strongly affects the overall conformational flexibility of macrolides and that the folded-in conformation occurs only with the introduction of sugar moieties at C3 and C5 positions.

Both conformations were reported to exist when macrolides bind to *H. marismortui* 50S subunit [52]. However, the existence of the folded-in conformation in the bound state has been questioned because a lower energy folded-out conformation of erythromycin exists in the crystal structure of the free compound, which was also observed for *E.coli* macrolide bound state [100].

NMR and molecular modeling have been used to predict the major conformations of telithromycin and RU 72366 in solution [101]. They are semisynthetic compounds produced from erythromycin lacking the cladinose sugar which was replaced by a keto group. They also possess a carbamate moiety at positions 11, 12 and alkyl-aryl side-chains. Different 2D heteronuclear NMR correlation data and long-range $^3J_{CH}$ coupling constants together with NOE experiments have been used in conjunction with molecular dynamics simulations to obtained conformations in solution which were found to be similar to those observed in the crystalline state. Authors concluded that owing to the presence of a keto group at position-3 the flexible right hand side of the macrocycle ring exhibits some deviations from the folded-out conformation found in related 14- and 15-membered macrolides. They also observed somewhat different orientation of the desosamine sugar with respect to that found in erythromycin. NOE correlations analysis did not show interesting contacts between lactone ring and imidazolyl-pyridine moiety in telithromycin indicating a substantial chain flexibility. On the other hand for RU 72366 NOE connectivities were observed suggesting that the chain is pushed back toward the right part of the macrocycle in the C-2 to C-6 region.

The three-dimensional solution-state structure of the macrolide antibiotic oxolide, an epoxy derivative of erythromycin, has also been investigated by a combination of NMR data and constrained molecular mechanics calculations [102]. Although $^3J_{H2,H3}$ coupling constants values measured in $CDCl_3$ and D_2O resemble those observed for the folded-out conformation, the characteristic NOE cross peaks indicative of the folded-out conformations were not detected in the ROESY spectrum. Furthermore, NOEs usually found for the folded-in conformation were not present as well. The effect of the epoxy moiety on the overall conformation of the macrolactone ring is the most probable cause for conformational differences. Stereochemistry and three-dimensional structure of anhydroerythromycin A, the degradation product of erythromycin A was determined in solution and solid state. The results indicated certain conformational flexibility in solution. It was suggested that for the flexible macrolide compounds two or more force fields should be included in computations for detailed conformational analysis [65].

A stereoisomer library of novel 14-membered macrolides has recently been constructed in an attempt to correlate stereochemistry and biological activity [103]. The superimposition of the most stable conformers indicated wide stereochemical diversity. The synthesis of a new macrolide has been performed and its conformation assessed by using NMR spectroscopy. The most stable structure calculated by molecular modeling was compared to the one obtained by NMR and two structures were found to be very similar.

There have been attempts to determine the three-dimensional structure of tylosin in aprotic solvents using molecular modeling and different force fields [104] and the obtained structures were not consistant with the one obtained in the crystal structure [41]. To better understand the conformation of tylosin in water solution Arsić et al. [105] performed a detailed study by using NMR and unconstrained molecular modeling methods. They concluded that tylosin neither adopted the folded-in nor the folded-out conformation probably due to the fact that the double bonds impose rigidity of some parts in the molecule with restricted conformational flexibility. The superimposition of the free state global minimum conformation on the crystal structure of tylosin bound to *H. marismortui* ribosomes revealed that the two structures resemble.

A comprehensive review on biosynthesis, structure, chemical properties and mode of action of 16-membered macrolides, with special emphasis on tylosin A and josamycin were recently published [106].

Conformation and relative configuration of the two novel 16-membered macrolide antibiotics, thuggacin A and B isolated from the myxobacterium *Sorangium cellulosum* and thuggacin C, a rearrangement product of thuggacin A, were studied in solution. NMR data including NOEs and vicinal coupling constants provided a reliable data for constructing a structural model. The conformation and relative configuration of thuggacins was calculated by using semi empirical PM3 calculations taking into account NOE constraints [92]. The structure of several hydroxy-aminoalkyl derivatives of 16-membered macrolide-josamycin have been evaluated by one- and two-dimensional NMR methods and FT-IR [107]. The lowest-energy structures of these compounds have been calculated and visualized by PM5 method at semi-empirical level of theory, taking into account spectroscopic data. For two josamycin derivatives equilibrium between two different structures were detected. PM5 calculations showed that the structures are stabilized by rather weak intramolecular hydrogen bonds, being in agreement with the spectroscopic data.

Two new 16-membered macrolides, tianchimycin A and B, were subjected to molecular mechanics calculations to search for the global energy minimum conformations [108]. Calculated structures were consistent with the ROESY correlations obtained in solution.

Hence, most of the conformational studies in solution of macrolide compounds have pointed towards their structural similarities and relative rigidity of the macrolactone ring with some flexibility noticed in certain regions. This might play a role in binding of these compounds to biological receptors.

References

[1] McGuire JM, Bunch RL, Anderson RC, Boaz HE, Flynn EH, Powell HM, et al. Ilotycin, a new antibiotic. Antibiot Chemother (Northfield) 1952, 2, 281–283.
[2] Flynn EH, Sigal MV Jr, Wiley PF, Gerzon K, Erythromycin. I. Properties and degradation studies. J Am Chem Soc 1954, 76, 3121–3131.
[3] Harris DR, McGeachin SG, Mills HH. The structure and stereochemistry of erythromycin A. Tetrahedron Lett 1965, 679–685.
[4] Pettinga CW, Stark WM, Van Abeele FR. The isolation of a second crystalline antibiotic from *Streptomyces erythreus*. J Am Chem Soc 1954, 76, 569–571.
[5] Wiley PF, Gale R, Pettinga CW, Gerzon K. Erythromycin. XII. 1 The isolation, properties and partial structure of erythromycin C. J Am Chem Soc 1957, 79, 6074–6077.
[6] Majer J, Martin JR, Egan RS, Corcoran JW. Antibiotic glycosides. 8. Erythromycin D, a new macrolide antibiotic. J Am Chem Soc 1977, 99, 1620–1622.
[7] Mikami Y, Yazawa K, Nemoto A, Komaki H, Tanaka Y, Gräfe U. Production of erythromycin E by pathogenic *Nocardia brasiliensis*. J Antibiot (Tokyo) 1999, 52, 201–202.
[8] Martin JR, DeVault RL, Sinclair AC, Stanaszek RS, Johnson P. A new naturally occurring erythromycin: erythromycin F. J Antibiot (Tokyo) 1982, 35, 426–430.
[9] Cevallos A, Guerriero A. Isolation and structure of a new macrolide antibiotic, erythromycin G, and a related biosynthetic intermediate from a culture of *Saccharopolyspora erythraea*. J Antibiot (Tokyo) 2003, 56, 280–288.

[10] Seiple IB, Zhang Z, Jakubec P, Langlois-Mercier A, Wright PM, Hog DT, et al. A platform for the discovery of new macrolide antibiotics. Nature 2016, 533, 338–345.

[11] Morimoto S, Takahashi Y, Watanabe Y, Omura S. Chemical modification of erythromycins. I. Synthesis and antibacterial activity of 6-*O*-methylerythromycins A. J Antibiot (Tokyo) 1984, 37, 187–189.

[12] Retsema J, Girard A, Schelkly W, Manousos M, Anderson M, Bright G, et al. Spectrum and mode of action of azithromycin (CP-62,993), a new 15-membered-ring macrolide with improved potency against Gram-negative organisms. Antimicrob Agents Chemother 1987, 31, 1939–1947.

[13] Jones RN, Biedenbach DJ. Antimicrobial activity of RU-66647, a new ketolide. Diagn Microbiol Infect Dis 1997, 27, 7–12.

[14] Jorgensen JH, Redding JS, Howell AW. *In vitro* activity of the new macrolide antibiotic roxithromycin (RU 28965) against clinical isolates of *Haemophilus influenzae*. Antimicrob Agents Chemother 1986, 29, 921–922.

[15] Atkins PJ, Herbert TO, Jones NB. Kinetic studies on the decomposition of erythromycin A in aqueous acidic and neutral buffers. Int J Pharm 1986, 30, 199–207.

[16] Cachet Th, Van den Mooter G, Hauchecorne R, Vinckier C, Hoogmartens J. Decomposition kinetics of erythromycin A in acidic aqueous solutions. Int J Pharm 1989, 55, 59–65.

[17] Barber J, Gyi JI, Lian L, Morris GA, Pye DA, Sutherland JK. The structure of erythromycin A in [²H₆] DMSO and buffered D₂O: full assignments of the ¹H and ¹³C NMR spectra. J Chem Soc, Perkin Transactions 2 1991, 1489–1494.

[18] Alam P, Buxton PC, Parkinson JA, Barber J. Structural studies on erythromycin A enol ether: full assignments of the ¹H and ¹³C NMR spectra. J Chem Soc, Perkin Transactions 2 1995, 1163–1167.

[19] Alam P, Embrey KJ, Parkinson JA, Barber J. Assignments of the ¹H and ¹³C NMR spectra of anhydroerythromycin A in organic and aqueous solutions. Magn Reson Chem 1996, 34, 837–839.

[20] Hassanzadeh A, Helliwell M, Barber J. Determination of the stereochemistry of anhydroerythromycin A, the principal degradation product of the antibiotic erythromycin A. Organic Biomol Chem 2006, 4, 1014–1019.

[21] Hassanzadeh A, Gorry PA, Morris GA, Barber J. Pediatric erythromycins: a comparison of the properties of erythromycins A and B 2′-ethyl succinates. J Med Chem 2006, 49, 6334–6342.

[22] Morris GA, Barjat H, Home TJ. Reference deconvolution methods. Prog Nucl Magn Reson Spectrosc 1997, 31, 197–257.

[23] Hassanzadeh A, Barber J, Morris GA, Gorry PA. Mechanism for the degradation of erythromycin A and erythromycin A 2′-ethyl succinate in acidic aqueous solution. J Phys Chem A 2007, 111, 10098–10104.

[24] Volmer DA, Hui JPM. Study of erythromycin A decomposition products in aqueous solution by solid-phase microextraction/liquid chromatography/mass spectrometry. Rapid Commun Mass Spectrom 1998, 12, 123–129.

[25] Morimoto S, Misawa Y, Asaka T, Kondah H, Watanabe Y. Chemical modification of erythromycins. Structure and antibacterial activity of acid degradation products of 6-*O*-methylerythromycin A. J Antibiot 1990, 43, 570–573.

[26] Nakagawa Y, Itai S, Yoshida T, Nagai T. Physicochemical properties and stability in the acid solution of a new macrolide antibiotic clarithromycin, in a comparison with erythromycin. Chem Pharm Bull 1992, 40, 725–728.

[27] Mordi MN, Pelta MD, Boote V, Morris GA, Barber J. Acid-catalyzed degradation of clarithromycin and erythromycin B: a comparative study using NMR spectroscopy. J Med Chem 2000, 43, 467–474.

[28] Awan A, Barber J, Brennan RJ, Parkinson JA. Structural studies on clarithromycin (6-*O*-methylerythromycin A): Assignments of the ¹H and ¹³C NMR spectra in organic and aqueous solutions. Magn Reson Chem 1992, 30, 1241–1246.

[29] Traunmüller F, Gattringer R, Zeitlinger MA, Graninger W, Müller M, Joukhadar C. Determination of telithromycin in human plasma and microdialysates by high-performance liquid chromatography. J Chromatogr B Analyt Technol Biomed Life Sci 2005, 822, 133–136.

[30] Wilhelm JM, Olenick NL, Corcoran JW. Interaction of antibiotics with ribosomes: Structure-function relationships and a possible common mechanism for the antibacterial action of the macrolides and lincomycin. Antimicrob Agents Chemother 1967, 10, 236–250.

[31] Kibwage IO, Hoogmartens J, Roets E, Vanderhaeghe H, Verbist L, Dubost M, et al. Antibacterial activities of erythromycins A, B, C and D and some of their derivatives. Antimicrob Agents Chemother 1985, 28, 630–633.

[32] Bhadra PK, Morris GA, Barber J. Design, synthesis, and evaluation of stable and taste-free erythromycin proprodrugs. J Med Chem 2005, 48, 3878–3884.

[33] Bhadra PK, Hassanzadeh A, Arsic B, Allison DG, Morris GA, Barber J. Enhancement of the properties of a drug by mono-deuteriation: reduction of acid-catalysed formation of a gut-motilide enol ether from 8-*deuterio*-erythromycin B. Org Biomol Chem 2016, 14, 6289–6296.

[34] Tsuzuki K, Sunazuka T, Marui S, Toyoda H, Omura S, Inatomi N, et al. Motilides, macrolides with gastrointestinal motor stimulating activity. I. *O*-substituted and tertiary *N*-substituted derivatives of 8,9-anhydroerythromycin A 6,9-hemiacetal. Chem Pharm Bull 1989, 37, 2687–2700.

[35] Sunazuka T, Tsuzuki K, Marui S, Toyoda H, Omura S, Inatomi N, et al. Motilides, macrolides with gastrointestinal motor stimulating activity. II. Quaternary *N*-substituted derivatives of 8,9-anhydroerythromycin A 6,9-hemiacetal and 9,9-dihydroerythromycin A 6,9-epoxide. Chem Pharm Bull 1989, 37, 2687–2700.

[36] Steinmetz WE, Shapiro BL, Roberts JJ. The structure of erythromycin enol ether as a model for its activity as a motilide. J Med Chem 2002, 45, 4899–4902.

[37] Morimoto S, Adachi T, Misawa Y, Nagate T, Watanabe Y, Omura S. Chemical modification of erythromycins IV. Synthesis and biological properties of 6-*O*-methylerythromycin B. J Antibiot 1990, 43, 544–549.

[38] Berisio R, Harms J, Schluenzen F, Zarivach R, Hansen HA, Fucini P, et al. A. Structural insight into the antibiotic action of telithromycin against resistant mutants. J Bacteriol 2003, 185, 4276–4279.

[39] Harms JM, Schlunzen F, Fucini P, Bartels H, Yonath A. Alterations at the peptidyl transferase centre of the ribosome induced by the synergistic action of the streptogramins dalfopristin and quinupristin. BMC Biol 2004, 2, 4–13.

[40] Schlunzen F, Zarivach R, Harms J, Bashan A, Tocilj A, Albrecht R, et al. Structural basis for the interaction of antibiotics with the peptidyl transferase centre in eubacteria. Nature 2001, 413, 814–821.

[41] Hansen JL, Ippolito JA, Ban N, Nissen P, Moore PB, Steitz TA. The structures of four macrolide antibiotics bound to the large ribosomal subunit. Mol Cell 2002, 10, 117–128.

[42] Hansen JL, Moore PB, Steitz TA. Structures of five antibiotics bound at the peptidyl transferase center of the large ribosomal subunit. J Mol Biol 2003, 330, 1061–1075.

[43] Tu D, Blaha G, Moore PB, Steitz TA. Structures of MLS$_B$K antibiotics bound to mutated large ribosomal subunits provide a structural explanation for resistance. Cell 2005, 121, 257–270.

[44] Stephenson GA, Stowell JG, Toma PH, Pfeiffer RR, Byrn SR. Solid-state investigations of erythromycin A dehydrate: structure, NMR spectroscopy, and hygroscopicity. J Pharm Sci 1997, 86, 1239–1244.

[45] Tenson T, Lovmar M, Ehrenberg M. The mechanism of action of macrolides, lincosamides and streptogramin B reveals the nascent peptide exit path in the ribosome. J Mol Biol 2003, 330, 1005–1014.

[46] Vazquez D. III. Mechanisms of action of antimicrobial and antitumor agents. In Concoran J. W. & Hahn F. E., Eds. Antibiotics, Berlin, Heidelberg, New York, Springer, 1975, 459–479.

[47] Nissen P, Hansen J, Ban N, Moore PB, Steitz TA. The structural basis of ribosome activity in peptide bond synthesis. Science 2000, 289, 920–930.

[48] Steinmetz WE, Bersch R, Towson J, Pesiri D. The conformation of 6-methoxyerythromycin A in water determined by proton NMR spectroscopy. J Med Chem 1992, 35, 4842–4845.

[49] Hansen LH, Mauvais P, Douthwaite S. The macrolide-ketolide antibiotic binding site is formed by structures in domains II and V of 23S ribosomal RNA. Mol Microbiol 1999, 31, 623–631.

[50] Mao JC-H, Puttermann M. The intermolecular complex of erythromycin and ribosome. J Mol Biol 1969, 44, 347–361.

[51] Saarma U, Spahn CMT, Nierhaus KH, Remme J. Mutational analysis of the donor substrate binding site of the ribosomal peptidyltransferase center. RNA 1998, 4, 189–194.

[52] Schlunzen F, Harms JM, Franceschi F, Hansen HAS, Bartels H, Zarivach R, et al. Structural basis for the antibiotic activity of ketolides and azalides. Structure 2003, 11, 329–338.

[53] Bulkley D, Innis CA, Blaha G, Steitz TA. Revisiting the structures of several antibiotics bound to the bacterial ribosome. PNAS 2010, 107(40), 17158–17163.

[54] Goldman RC, Fesik SW, Doran CC. Role of protonated and neutral forms of macrolides in binding to ribosomes from Gram-positive and Gram-negative bacteria. Antimicrob Agents Chemother 1990, 34, 426–431.

[55] Sabath LD, Lorian V, Gerstein D, Loder PB, Finland M. Enhancing effect on [slc] alkalinisation of the medium on the activity of erythromycin against gram-negative bacteria. Appl Microbiol 1968, 16, 1288–1292.

[56] Omura S, Namiki S, Shibata M, Muro T, Sawada J. Studies on the antibiotics from *Streptomyces spinichromogenes var. kujimyceticus*. V. Some antimicrobial characteristics of kujimycin A and kujimycin B against macrolide resistant staphylococci. J Antibiot 1970, 23, 448–460.

[57] Dette GA, Knothe H, Kaula S. Modifying effects of pH and temperature on (^{14}C) erythromycin uptake into *Staphylococcus aureus*-relation to antimicrobial activity. Zentralbl Bakteriol Parasitenkd Infektionskr Hyg Abt 1 Orig Reihe A 1987, 265, 393–403.

[58] Pestka S, Lemahieu R, Miller P. Correlation of effects of erythromycin analogues on intact bacteria and on [^{14}C] erythromycin binding to *Escherichia coli* ribosomes. Antimicrob Agents Chemother 1974, 6, 489–491.

[59] Fernandez-Munoz R, Vasquez D. Quantitative binding of MC-erythromycin to *E. coli* ribosomes. J Antibiot 1973, 26, 107–108.

[60] Pestka S. Binding of [14C] erythromycin to *Escherichia coli* ribosomes. Antimicrob Agents Chemother 1974, 6, 474–478.

[61] Barber J, Gyi JI, Pye DA. Specific weak binding of erythromycin A (ketone) and chloramphenicol to 50S subunits of *Escherichia coli* ribosomes. A 1H-NMR study. J Chem Soc Chem Commun 1991, 1249–1252.

[62] Pye DA, Gyi JI, Barber J. Tautomeric recognition of erythromycin A by ribosomes: A ^{1}H nuclear magnetic resonance study. J Chem Soc Chem Commun 1990, 1143–1145.

[63] Arsic B, Awan A, Brennan RJ, Aguilar JA, Ledder R, McBain AJ, et al. Theoretical and experimental investigation on clarithromycin, erythromycin A and azithromycin and descladinosyl derivatives of clarithromycin and azithromycin with 3-*O* substitution as antibacterial agents. Med Chem Commun 2014, 5, 1347–1354.

[64] Bertho G, Gharbi-Benarous J, Delaforge M, Girault JP. Transferred nuclear Overhauser effect study of macrolide-ribosome interactions: correlation between antibiotic activities and bound conformations. Bioorg Med Chem 1998, 6, 209–221.

[65] Kosol S, Schrank E, Bukvić-Krajačić M, Wagner G, Meyer NH, et al. Probing the interactions of macrolide antibiotics with membrane-mimetics by NMR spectroscopy. J Med Chem 2012, 55, 5632–5636.

[66] Shinde PB, Han AR, Cho J, Lee SR, Ban YH, Yoo YJ, et al. Combinatorial biosynthesis and antibacterial evaluation of glycosylated derivatives of 12-membered macrolide antibiotic YC-17. J Biotechnol 2013, 168, 142–148.

[67] Nakano H, Sugawara A, Hirose T, Gouda H, Hirono S, Omura S, et al. An architectonic macrolide library based on a C2-symmetric macrodiolide toward pharmaceutical compositions. Tetrahedron 2014, 71, 6569–6579.

[68] Glanzer S, Pulido SA, Tutz S, Wagner GE, Kriechbaum M, Gubensäk N, et al. Structural and functional implications of the interaction between macrolide antibiotics and bile acids. Chem Eur J 2015, 21, 1–10.

[69] Chelvan P, Hamilton-Miller J, Brumfitt W. Biliary excretion of erythromycin after parenteral administration. Br J Clin Pharmacol 1979, 8, 233–235.

[70] Dinos GP, Connell SR, Nierhaus KH, Kapaxis DL. Erythromycin, roxithromycin, and clarithromycin: use of slow-binding kinetics to compare their *in vitro* interaction with a bacterial ribosomal complex active in peptide bond formation. Mol Pharmacol 2003, 63, 617–623.

[71] Garza-Ramos G, Xiong L, Zhong P, Mankin A. Binding site of macrolide antibiotics on the ribosome: new resistance mutation identifies a specific interaction of ketolides with rRNA. J Bacteriol 2001, 183, 6898–6907.

[72] Douthwaite S, Aagaard C. Erythromycin binding is reduced in ribosomes with conformational alterations in the 23S rRNA peptidyl transferase loop. J Mol Biol 1993, 232, 725–731.

[73] Kouvela EC, Kalpaxis DL, Wilson DN, Dinos GP. Distinct mode of interaction of a novel ketolide antibiotic that displays enhanced antimicrobial activity. Antimicrob Agents Chemother 2009, 53, 1411–1419.

[74] Morimura T, Hashiba M, Kameda H, Takami M, Takahama H, Ohshige M, et al. Identification of macrolide antibiotic-binding human_p8 protein. J Antibiot 2008, 61, 291–296.

[75] Xiong L, Korkhin Y, Mankin AS. Binding site of the bridged macrolides in the *Escherichia coli* ribosome. Antimicrob Agents Chemother 2005, 49, 281–288.

[76] Yan K, Hunt E, Berge J, May E, Copeland RA, Gontarek RR. Fluorescence polarization method to characterize macrolide-ribosome interactions. Antimicrob Agents Chemother 2005, 49, 3367–3372.

[77] Brandt-Rauf P, Vince R, LeMahieu R, Pestra S. Fluorescent assay for estimating the binding of erythromycin derivatives to ribosomes. Antimicrob Agents Chemother 1978, 14, 88–94.

[78] Giambattista MD, Engelborghs Y, Nysen E, Cocito C. Kinetics of binding of macrolides, lincosamides, and synergimycins to ribosomes. J Biol Chem 1987, 262, 8591–8597.

[79] Everett JR, Tyler JW. The conformational analysis of erythromycin A. J Chem Soc Perkin Trans. 1987, 2, 1659–1667.

[80] Woodward RB. Structure and biogenesis of the macrolides. Angew Chem 1957, 69, 50–58.

[81] Luger P, Maier R. Molecular structure of 9-deoxy-11-deoxy-9-11-(imino(2-(2-methoxyethoxy) ethylidene)oxy)-(9S)-erythromycin, a new erythromycin derivative. J Cryst Mol Struct 1979, 9, 329–338.

[82] Miroshnyk I, Mirza S, Zorky PM, Heinämäki J, Yli-Kauhaluoma J, Yliruusi J. A new insight into solid-state conformation of macrolide antibiotics. Bioorg Med Chem 2008, 16, 232–239.

[83] Lazarevski G, Vinković M, Kobrehel G, Đokić S. Conformational analysis of azithromycin by nuclear magnetic resonance spectroscopy and molecular modelling. Tetrahedron 1993, 49, 721–730.

[84] Gharbi-Benarous J, Ladam P, Delaforge M, Girault J-P. Conformational analysis of major metabolites of macrolide antibiotics roxithromycin and erythromycin A with different biological properties by NMR spectroscopy and molecular dynamics. J Chem Soc Perkin Trans 2, 1993, 2303–2321.

[85] Awan A, Brennan RJ, Regan AC, Barber J. Conformational analysis of the erythromycin analogues azithromycin and clarithromycin in aqueous solution and bound to bacterial ribosomes. J Chem Soc Chem Commun 1995, 1653–1654.

[86] Awan A, Brennan RJ, Regan AC, Barber J. The conformations of the macrolide antibiotics erythromycin A, azithromycin and clarithromycin in aqueous solution: a ^{1}H NMR study. J Chem Soc Perkin Trans 2, 2000, 1645–1652.

[87] Steinmetz WE, Sadowsky JD, Rice JS, Roberts JJ, Bui YK. Determination of the aqueous-phase structure of 6-*O*-methylerythromycin from NMR constraints. Magn Reson Chem 2001, 39, 163–172.

[88] Košutić-Hulita N, Matak-Vinković D, Vinković M, Novak P, Kobrehel G, Lazarevski G. Conformational behaviour of 11-*O*-methylazithromycin in the solid and solution state. Croat Chem Acta 2001, 74, 327–341.

[89] Novak P, Banić Tomišić Z, Tepeš P, Lazarevski G, Plavec J, Turkalj G. Conformational analysis of oleandomycin and its 8-methylene-9-oxime derivative by NMR and molecular modelling. Org Biomol Chem 2005, 3, 39–47.

[90] Novak P, Barber J, Čikoš A, Arsić B, Plavec J, Lazarevski G, et al. Free and bound state structures of 6-*O*-methyl homoerithromycins and epitope mapping of their interactions with ribosomes. Bioorg Med Chem 2009, 17, 5857–5867.

[91] Novak P, Tatić I, Tepeš P, Koštrun S, Barber J. Systematic approach to understanding macrolide-ribosome interactions: NMR and modelling studies of oleandomycin and its derivatives. J Phys Chem A 2006, 110, 572–579.

[92] Steinmetz H, Irschik H, Kunze B, Reichenbach H, Höfle G, Jansen R. Thuggacins, macrolide antibiotics active against *Mycobacterium tuberculosis*: isolation from myxobacteria, structure elucidation, conformation analysis and biosynthesis. Chem Eur J 2007, 13, 5822–5832.

[93] Kamenar B, Mrvoš-Sermek D, Vicković I, Nagl A. Structure of 10,10-dihydro-10-deoxo-10a-methyl-10a-aza-10a-homoerythronolide A. Acta Crystallogr Sect C 1990, 46, 1964–1966.

[94] Kamenar B, Mrvoš-Sermek D, Banić Z, Nagl A, Kobrehel G. Crystal and molecular structure of 9-deoxo-9a-ethyl-9a-aza-9a-homoerythronolide A. Croat Chem Acta 1991, 64, 153–160.

[95] Sheldrick GM, Kojić-Prodić B, Banić Z, Kobrehel G, Kujundžić N. Structure of 9-deoxo-9a-*N*-[*N'*-(4-pyridyl)-carbamoyl]-9a-aza-9a-homoerythromycin A and conformational analysis of analogous 9a-aza 15-membered azalides in the solid state. Acta Crystallogr Sect B 1995, 51, 358–366.

[96] Lazarevski G, Vinković M, Kobrehel G, Đokić S, Metelko B. Conformational analysis of 9-deoxo-9a-aza-9a- and 9-deoxo-8a-aza-8a-homoerythromycin A 6,9-cyclic iminoethers. Tetrahedron 1994, 50, 12201–12210.

[97] Bukvić Krajačić M, Dumić M, Novak P, Cindrić M, Koštrun S, Fajdetić A, et al. Discovery of novel ureas and thioureas of 3-decladinosyl-3-hydroxy 15-membered azalides active against efflux-mediated resistant *Streptococcus pneumoniae*. Bioorg Med Chem Lett 2011, 21, 853–856.

[98] Bertho G, Gharby-Benarous J, Delaforge M, Lang C, Parent A, Girault JP. Conformational analysis of ketolide, conformations of RU 004 in solution and bound to bacterial ribosomes. J Med Chem 1998, 41, 3373–3386.

[99] Ćaleta I, Čikoš A, Žiher D, Đilović I, Dukši M, Gembarovski D, et al. Synthesis, NMR and X-ray structure analysis of macrolide aglycons. Struct Chem 2012, 23, 1785–1796.

[100] Dunkle JA, Xiong L, Mankin AS, Cate JHD. Structures of the *Escherichia coli* ribosome with antibiotics bound near the peptidyl transferase center explain spectra of drug action. Proc Natl Acad Sci 2010, 107, 17152–17157.

[101] Evrard-Todeschi N, Gharbi-Benarous J, Gaillet C, Verdier L, Gertho G, Lang C, et al. Conformations in solution and bound to bacterial ribosomes of ketolides HMR 3647 (telithromycin) and RU 72366: a new class of highly potent antibacterials. Bioorg Med Chem 2000, 8, 1579–1597.

[102] Steinmetz WE, Sparrow A, Somsouk M. Determination of three-dimensional solution-phase structure of the macrolide antibiotic oxolide in CD_2Cl_2 and D_2O from NMR constraints. Magn Reson Chem 2005, 43, 16–20.

[103] Nakano H, Sugawara A, Hirose T, Gouda H, Hirono S, Omura S, et al. An architectonic macrolide library based on a C2-symmetric macrolide toward pharmaceutical composition. Tetrahedron 2015, 71, 6569–6579.

[104] Ivanov PM. CONFLEX/MM3 search/minimization study of the conformations of the macrolide antibiotic tylosin. J Mol Struct 2002, 606, 217–229.

[105] Arsic B, Aguilar JA, Bryce RA, Barber J. Conformational study of tylosin A in water and full assignments of ^{1}H and ^{13}C spectra of tylosin A in D$_2$O and tylosin B in CDCl$_3$. Magn Reson Chem 2017, 55, 367–373.

[106] Arsic B, Barber J, Čikoš A, Mladenovic M, Stankovic N, Novak P. 16-Membered macrolide antibiotics: a review. https://doi.org/10.1016/j.ijantimicag.2017.05.020.

[107] Przybylski P, Pyta K, Stefanska J, Brzezinski B, Bartl F. Structure elucidation, complete NMR assignment and PM5 theoretical studies of new hydroxy-aminoalkyl-α, β-unsaturated derivatives of the macrolide antibiotic josamycin. Magn Reson Chem 2010, 48, 286–296.

[108] Wang X, Tabudravu J, Jaspars M, Deng H. Tianchimycins A-B, 16-membered macrolides from the rare actinomycete *Saccharothrix xinjiangensis*. Tetrahedron 2013, 69, 6060–6064.

Goran Kragol

2 The semisynthetic routes towards better macrolide antibiotics

2.1 Introduction

Macrolide antibiotics are characterized by the presence of a 12- to 16-membered macrocyclic lactone ring to which one or more sugars are attached. Over 50 years ago, Erythromycin A, a natural product isolated from *Saccharopolyspora erythraea*, was the first macrolide to be introduced for clinical use (Fig. 2.1). Although other natural macrolides isolated form *Streptomyces spp.* (oleandomycin, josamycin and spiramycin) can also be found in clinical use, the erythromycin-based 14-membered macrolide scaffold is still the highly predominant template for further derivatization to obtain better macrolide antibiotics. This is not a surprise as erythromycin, as well as some erythromycin derivatives, is currently available in bulk quantities from a number of suppliers and therefore is a reasonably cheap starting material. Extensive chemical modifications of erythromycin A have led to the development of numerous semisynthetic derivatives with broader antimicrobial spectra, more favorable pharmacokinetic properties and better tolerability. In general, there are three generations of macrolide antibiotics derived from erythromycin A (Fig. 2.1): the first generation, which includes erythromycin and some simple derivatives, *i.e.* 2'-esters or lipophilic acid-addition salts; the second generation, which was designed to improve primarily pharmacokinetic properties and was commercially and clinically quite successful, *i.e.* roxithromycin (Rulid), clarithromycin (Clarith, Biaxin) and azithromycin (Sumamed, Zithromax); and the third generation, which has the difficult task of fighting against bacterial resistance and includes exclusively various ketolides, among which telithromycin (Ketek) is the only one on the market while others are still in development. Semisynthetic routes towards selected examples of macrolide antibiotics of the second and third generations will be described in this chapter. Currently, there is no the fourth generation in clinical study, but some promising novel platforms have been developed recently and their usefulness to produce new classes of macrolides will be examined.

2.2 Erythromycin

Erythromycin A (Fig. 2.1), the first clinically used 14-membered macrolide, is a natural fermentation product isolated from the culture broth of the soil-dwelling fungus *Saccharopolyspora erythrea* [1]. The structure of erythromycin consists of a 14-membered macrolactone ring, having 10 chiral centers and bearing three hydroxyls

Fidelta d.o.o., Prilaz b. Filipovica 29, 10000 Zagreb, Croatia

https://doi.org/10.1515/9783110515756-002

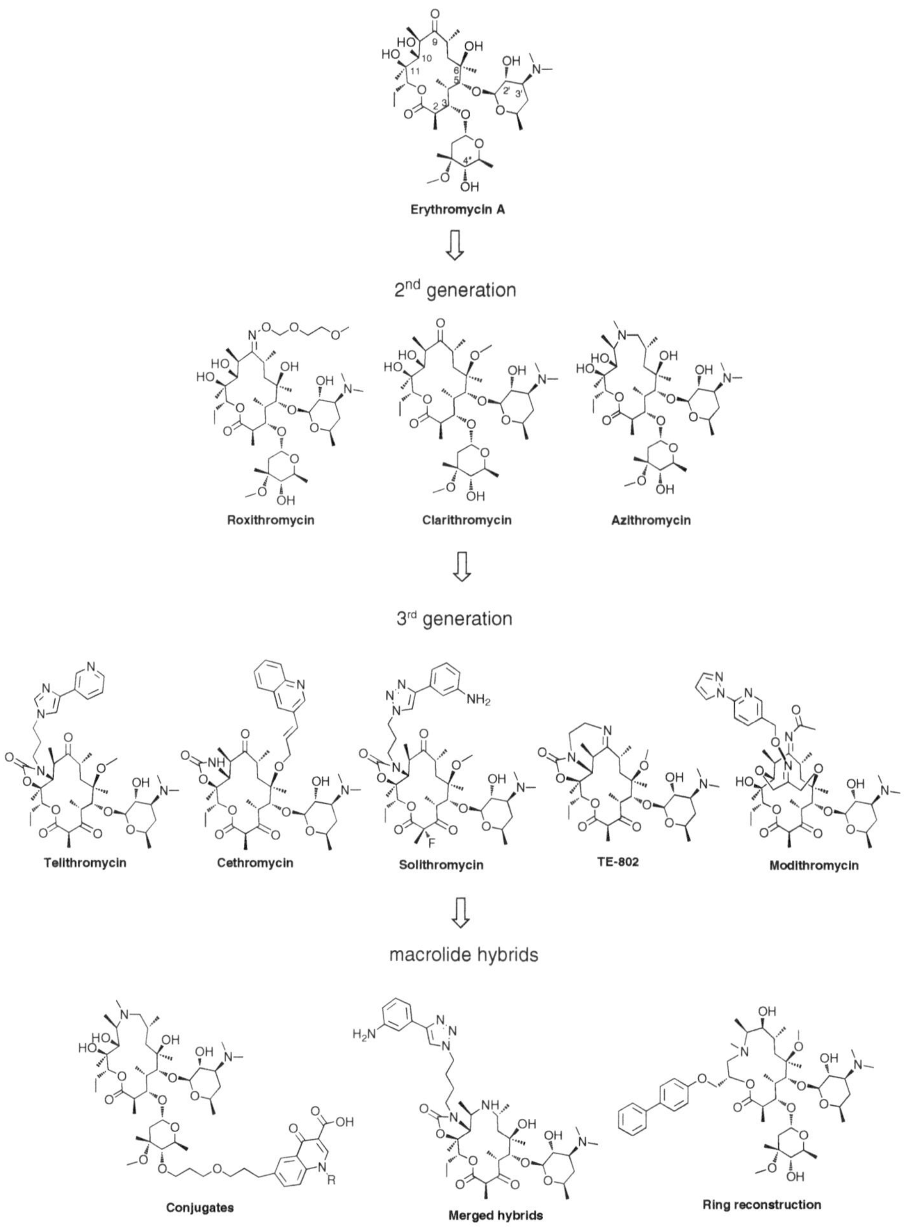

Fig. 2.1: Evolution of macrolide antibiotics.

and a keto group as well as six methyl groups and one ethyl group. At positions 3 and 5 (for the usual numbering of the erythromycin scaffold, see Fig. 2.1) of the macrolactone ring, two sugar moieties, cladinose and desosamine, respectively, are attached, further increasing the complexity and size of the erythromycin molecule. As such, erythromycin belongs to the useful drug classes that do not follow Lipinski's Rule-of-Five [2].

Large-scale production of erythromycin A still relies on the organism *Saccharopolyspora erythraea* (Fig. 2.2). The industrial production of erythromycin A started in the 1950s. The early fermentation process afforded about 0.25–1.0 g erythromycin per liter. After 60 years of improvement, yields have increased to around 10–13 g per liter. The fermentation process still delivers a mixture of erythromycin A as the major component (desirably >90%), contaminated with partially hydroxylated or methylated intermediates: erythromycins B, C and D (Fig. 2.2).

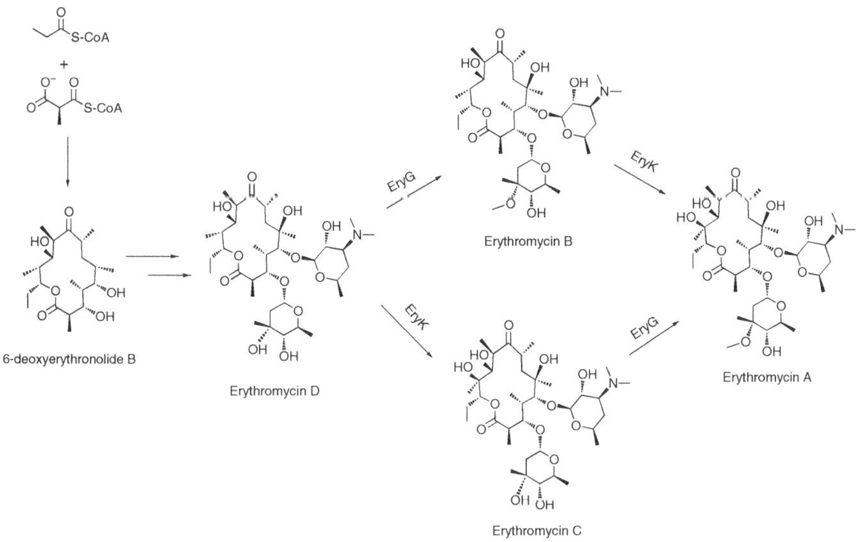

Fig. 2.2: Biosynthetic pathway and structures of erythromycin A and its intermediates B, C and D.

In addition to poor bioavailability and a short half-life, the main disadvantage of erythromycin A is its fast decomposition in the acidic conditions of stomach fluid. The mechanism of acidic decomposition involves intramolecular cyclization reactions, leading to the formation of anhydrohemiketal and spiroketal derivatives (Fig. 2.3) that are antibacterially inactive [3].

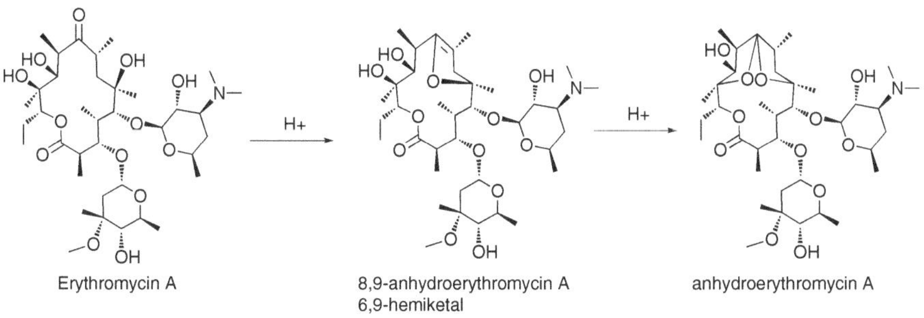

Fig. 2.3: Decomposition of erythromycin in acidic conditions.

2.3 Second generation of macrolide antibiotics

The discovery of semisynthetic second generation of macrolide antibiotics was essentially fueled by a desire to improve erythromycin's stability in acidic media. As acidic decomposition begins with the addition of the C6-hydroxy group to the C9-ketone (Fig. 2.3), researchers devised several successful solutions to block either 6-hydroxyl (6-OH) or 9-keto moiety. Each solution provided novel macrolide antibiotics that show improved pharmaceutical properties as well as an expanded antibacterial activity spectrum.

2.3.1 Roxithromycin

The transformation of C9-ketone to an oxime ether was shown to be a suitable method for the prevention of intramolecular cyclization reactions in acidic media [4]. Among the library of prepared oxime ethers, roxithromycin was chosen as the most promising compound. The synthetic procedure for the preparation of roxithromycin is quite straightforward, taking only two steps starting from erythromycin A (Scheme 2.1).

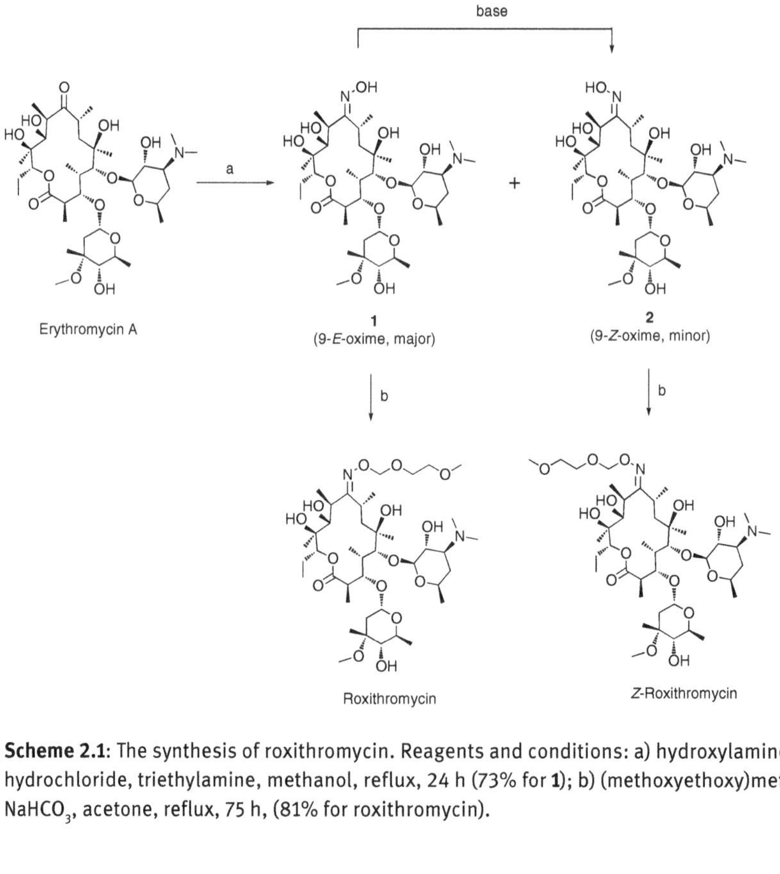

Scheme 2.1: The synthesis of roxithromycin. Reagents and conditions: a) hydroxylamine hydrochloride, triethylamine, methanol, reflux, 24 h (73% for **1**); b) (methoxyethoxy)methyl chloride, NaHCO$_3$, acetone, reflux, 75 h, (81% for roxithromycin).

The first reaction step involves formation of an oxime at position C9 of unprotected erythromycin A. The reaction of oxime formation is stereospecific, providing *E*-oxime **1** as the major product, contaminated with a small ratio of *Z*-isomer **2** that can be easily separated by crystallization. A hydroxy group of the oxime can be then selectively alkylated without a need for the protection of C2'-hydroxy (2'-OH) or C4"-hydroxy (4"-OH) moieties. Acidic stability of roxithromycin was moderate but substantially improved when compared to erythromycin A [4]. Base-promoted *E* to *Z* oxime isomerization provided a *Z*-isomer of roxithromycin that was clearly less active *in vivo* when compared to the *E*-isomer.

2.3.2 Clarithromycin

Another solution to decrease intramolecular ketalization of erythromycin A was blockage of the C6-hydroxy group by simple alkylation. However, the presence of five different hydroxy groups in the starting macrolide scaffold heavily complicates selective alkylation of only 6-OH. The first successful alkylation of 6-OH was accomplished on a 2'-*O*-,3'-*N*-bis(benzyloxycarbonyl)-*N*-demethyl erythromycin A (**3**, Scheme 2.2) [5]. The protection of 2'-OH was necessary because it is the most reactive hydroxyl of the erythromycin scaffold. The protection of 3'-amine blocks formation of quaternary ammonium salt during alkylations. The alkylation of **3** with methyl iodide in the presence

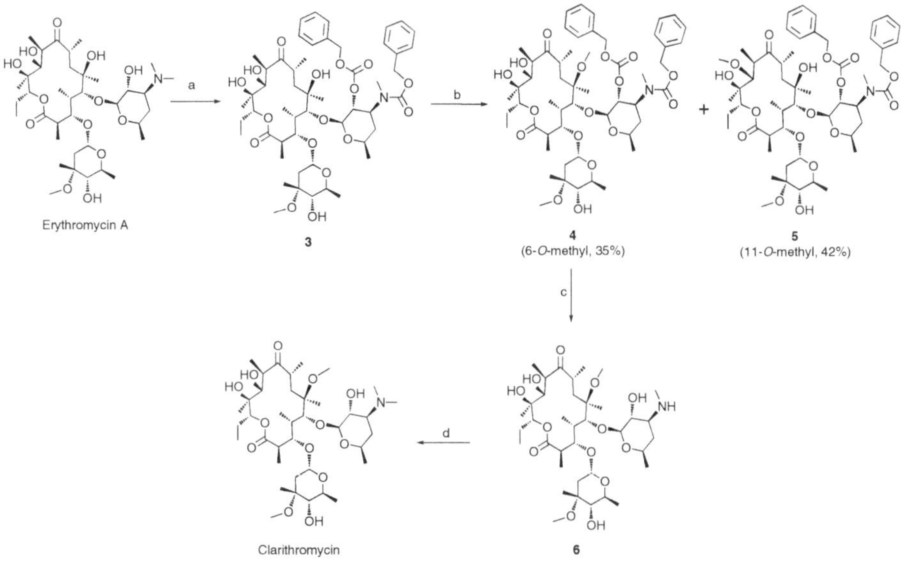

Scheme 2.2: The first synthesis of clarithromycin. Reagents and conditions: a) benzyl chloroformate, NaHCO$_3$, (60%); b) NaH, methyl iodide, DMSO/THF, 0 °C; c) H$_2$, Pd/C, ethanol, sodium acetate/acetic acid buffer (pH 5); d) H$_2$, formaldehyde (14% over three steps).

of sodium hydride provided a mixture of methylated products from which 6-O-methyl derivative **4** was isolated by chromatography. Subsequent catalytic hydrogenation to remove benzyloxycarbonyl (Cbz) protections afforded 3'-N-demethyl-6-O-methylerythromycin A (**6**). Reductive methylation of **6** to regain 3'-N-dimethylamino moiety provided clarithromycin in the best yield of 39% (contaminated with 6,11-di-O-methyl and 6,12-di-O-methyl derivatives) [6]. As expected, clarithromycin was quite stable in acidic conditions but also more active against bacteria when compared to erythromycin A.

Apart from desired 6-O-methyl derivative **4**, above alkylation procedure preferentially provides 11-O-methyl derivative **5** (42% yield). Further development of alkylation conditions revealed that the alkylation of suitable 2'-O-4"-O-protected erythromycin A 9-E-oxime **7**, provides preferentially 6-O-alkylated products in high yields (*i.e.* **10**, Scheme 2.3) [6]. The protection of 9-E-oxime was necessary to block the formation of O-alkyl oximes as the regeneration of the 9-keto group from alkylated oximes cannot be performed in mild conditions. Fortunately, it turned out that the bulkiness of the oxime protection group also determines regioselectivity and rate of the alkylation: bulkier protection affords better regioselectivity towards 6-O-alkylation but a slower reaction time. Eventually, 2-chlorobenzyl group oxime protection (**9**, Scheme 2.3) was chosen as the most suitable for the 6-OH alkylation studies. The optimal alkylation conditions involve DMSO/THF (1:1) as a solvent, KOH as a base and methyl iodide as a methylation agent. Subsequent steps consist of, firstly, catalytic hydrogenation to cleave 2-chlorobenzoyl oxime protection, secondly, catalytic hydrogenation to cleave benzyloxycarbonyl protections, and then, reductive methylation to regain 3'-N-dimethyl amino moiety. The obtained clarithromycin 9-E-oxime **12** was finally transformed to clarithromycin by conversion of an 9-E-oxime to a 9-keto moiety.

The synthesis of clarithromycin was further improved by ascertaining that bulky trimethylsilyl (TMS) protection of 2'-OH also effectively blocks the formation of tertiary amino salt at the 3' position (Scheme 2.4) [7]. Therefore, the use of toxic benzylchloroformate in large amounts is avoided as well as a two-step procedure for its catalytic removal and subsequent reductive methylation. The initial low regioselectivity of silylation was substantially improved if a mixture of TMS-chloride and TMS-imidazole was used to afford a 2',4"-O-bis-TMS derivative **14** in a 96% yield. Moreover, regioselectivity of the subsequent methylation of 6-OH was also improved to over 90%. The cleavage of 2-chlorobenzoyl and TMS groups was achieved by catalytic hydrogenation in one pot followed by regeneration of ketone at C9. The overall yield of clarithromycin from erythromycin A 9-E-oxime **1** was improved to 48% with only one purification of the intermediate needed.

Although quite remarkable progress in the semisynthesis of clarithromycin was achieved, the catalytic hydrogenation step was quite complicated and expensive to perform on a large scale. Therefore, the novel protections of 9-E-oxime **1** were eventually introduced. For example, isopropoxy cyclohexylketal protection of oxime hydroxyl can be used for large scale production of clarithromycin as it could be cleaved off together with 2'- and 4"-silyl protections in one pot under mild conditions (Scheme 2.5) [8].

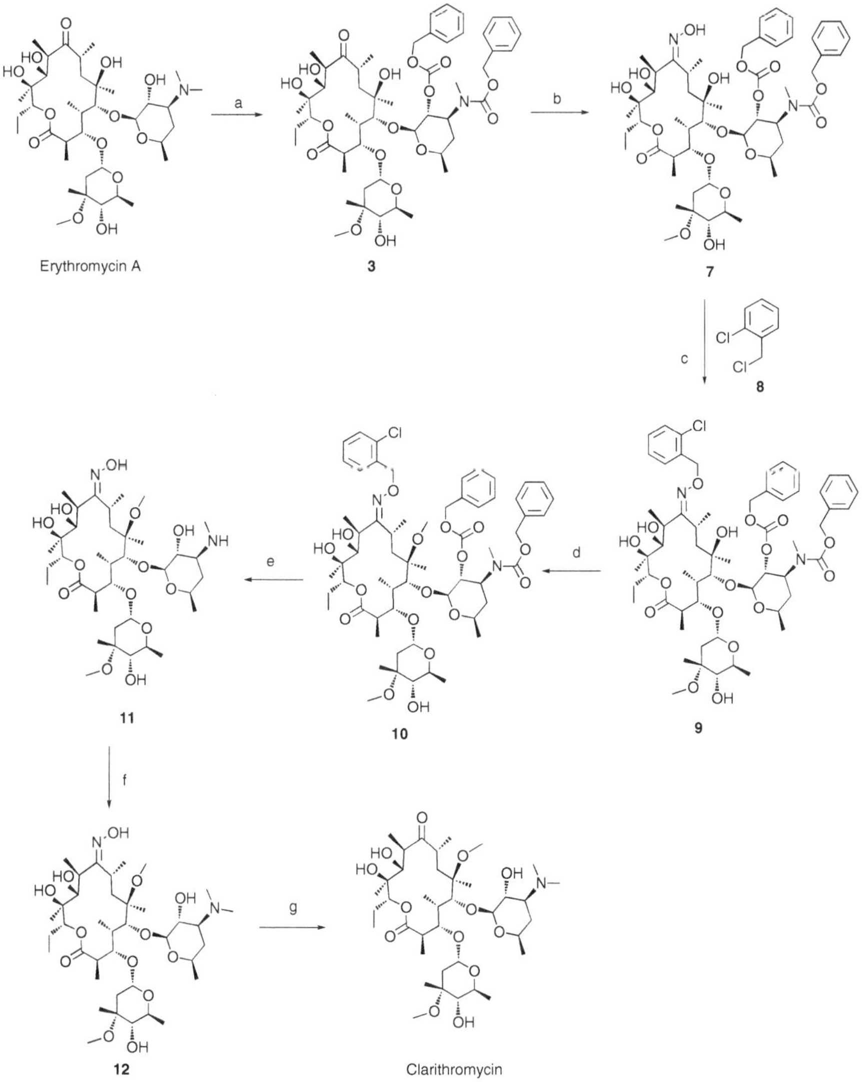

Scheme 2.3: Improved synthesis of clarithromycin. Reagents and conditions: a) benzyl chloroformate, NaHCO$_3$, (60%); b) NH$_2$OH x HCl, sodium acetate, methanol, rt (75%); c) **8**, KOH, DMF, 0 °C (99%); d) methyl iodide, KOH, DMSO/THF, 5 °C (86%); e) H$_2$, Pd/C, methanol/water, acetic acid (90%); f) HCOOH, HCHO, methanol, reflux (94%); g) NaHSO$_3$, ethanol/water, reflux (68%).

Interestingly, the above method was also used to prepare a library of various 6-*O*-alkyl derivatives [9]. Instead of KOH, potassium *tert*-butoxide (*t*BuOK) was used as a base to introduce less reactive alkyl halides. Although larger substituents at 6-OH are well tolerated, none of the prepared derivatives showed substantially better antibacterial activity in comparison to erythromycin A.

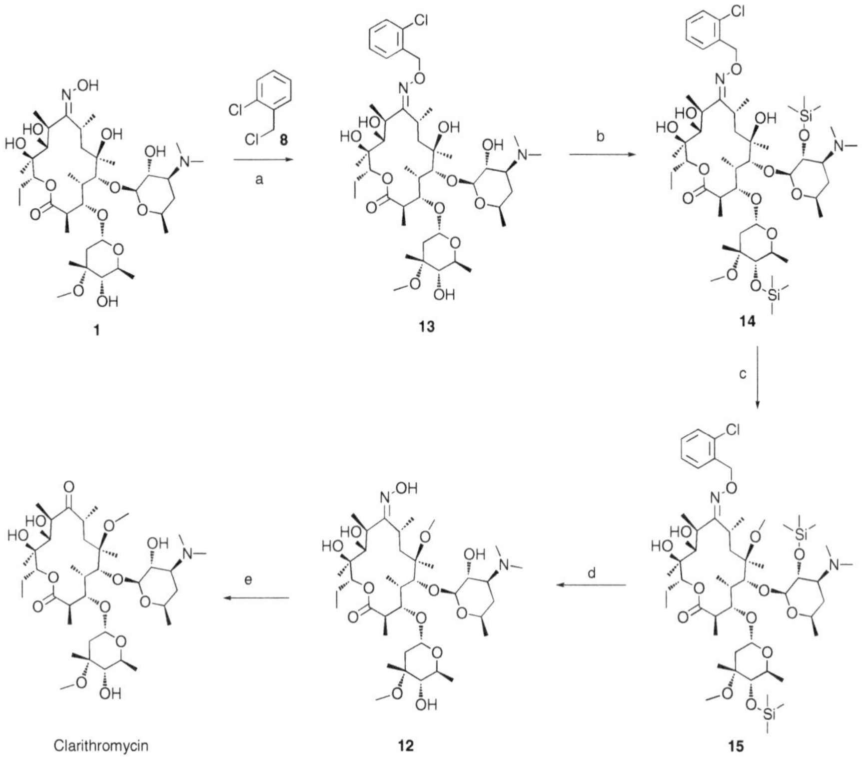

Scheme 2.4: Improved synthesis of clarithromycin. a) **8**, KOH, DMF, rt (89%); b) TMS-Cl, TMS-imidazole, ethyl acetate, rt (96%); c) methyl iodide, KOH, DMSO/THF, 5 °C (83%); d) Pd/C, ammonium formate/formic acid, methanol; e) NaHSO$_3$, ethanol/water, reflux (57% over two steps).

More recently, some new protective groups for the 9-*E*-oxime **1** were introduced, for example 2-pyrimidine [10] and tritylhydrazone [11] (**20** and **21**, respectively, Fig 2.4). It seems that regioselectivity and yields are comparable to isopropoxy cyclohexylketal protection but different protections allowed patents for a large scale synthesis of clarithromycin that did not infringe previous patents.

2.3.3 Azithromycin

Erythromycin A 9-*E*-oxime **1** also played a pivotal role in the discovery of a novel macrolide scaffold, termed azalide, in which the macrolactone ring is enlarged from 14- to 15-membered macrolides and contains one nitrogen atom. In this case, the usual Beckmann rearrangement reaction of cyclic oximes afforded an unusual reaction product when applied to oxime **1**. However, the surprising course of this reaction eventually led to the discovery of azithromycin, which is one of the best-selling

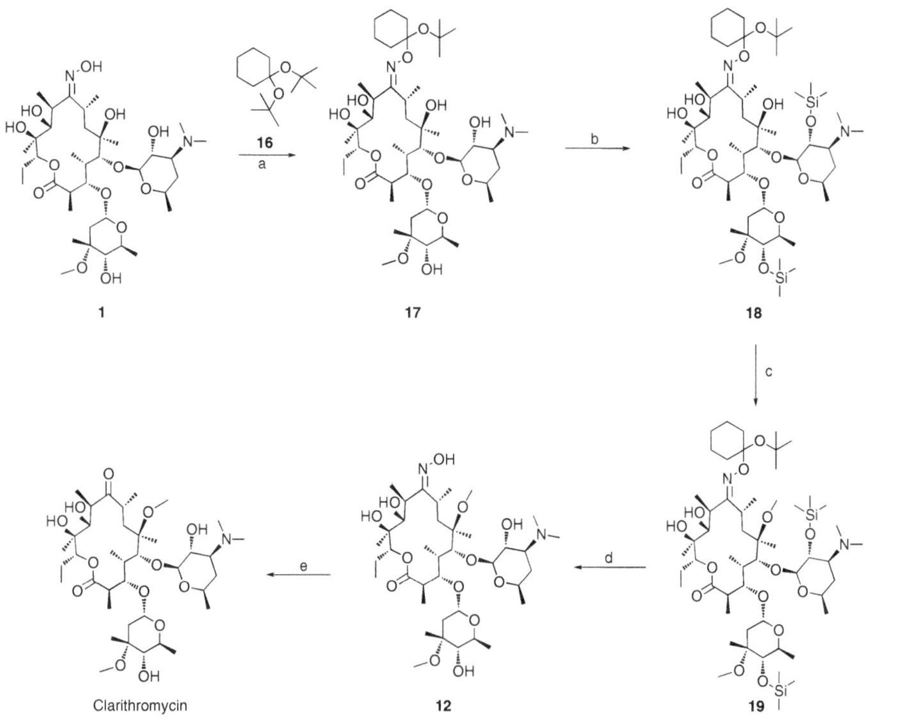

Scheme 2.5: Large scale preparation of clarithromycin. Reagents and conditions: a) **16**, pyridine hydrochloride, DCM, rt (85%); b) 1,1,1,3,3,3-hexamethyldisilazane, DMF, rt (73%); c) methyl iodide, KOH, DMSO/THF, rt (95%); d) HCOOH, ethanol/water, rt (96%); e) NaHSO₃, HCOOH, ethanol/water, reflux (77%).

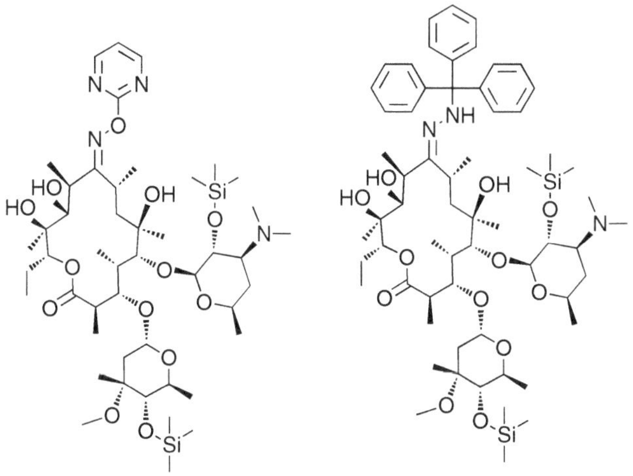

Fig 2.4: Recently published protection strategy for clarithromycin synthesis.

antibiotics in the world. In an attempt to introduce a nitrogen atom into the macrol-actone ring *via* a Beckmann rearrangement of erythromycin A 9-*E*-oxime **1**, a series of 9a-sulfonyloximes **22** was prepared (Scheme 2.6). Nevertheless, attempts to rearrange sulfonyl oximes **22** to a lactam **23** provided primarily erythromycin A-6,9-iminoether **24** instead (Scheme 2.6) [12].

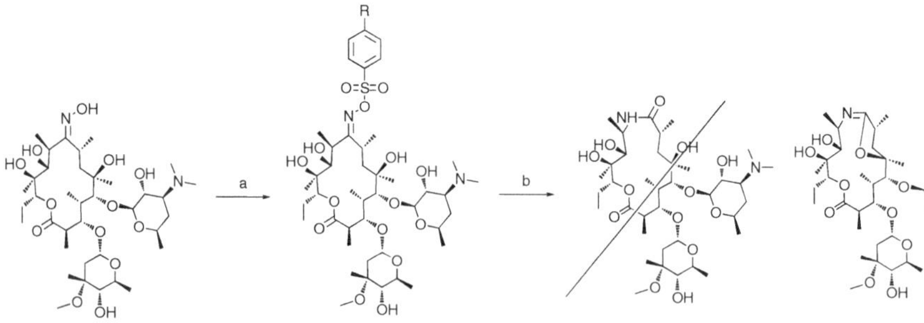

Scheme 2.6: Beckmann rearrangement of erythromycin A 9-*E*-oxime. Reagents and conditions: a) *p*-substituted benzenesulphonyl chloride, NaHCO₃, acetone; b) 2 N HCl, DCM.

The same product **24** was obtained during the Beckmann rearrangement of erythromycin A oxime **1** with tosyl chloride (TsCl) in aqueous acetone (Scheme 2.7). It was later found that the by-product of this reaction is *seco*-macrolide **25**, which is prob-ably formed by the hydrolysis of 6,9-iminoether **24** during the work-up process [13]. If the reaction is performed at a lower temperature (−45 °C), erythromycin A 9,11-iminoether **26** is also formed together with a small amount of lactam **23** [14]. Although isolated iminoethers **24** and **26** are stable as solids, they readily isomerize to each other in organic solvents. Nevertheless, reductions of both iminoethers provide the same product, 9-deoxo-9a-aza-9a-homoerythromycin (**27**). Two reduction methods were initially reported: catalytic hydrogenation in glacial acetic acid using platinum oxide provided ~80% yield of the amine **27**, while reduction with metal hydrides afforded a somewhat lower yield (~60%) due to cleavage of the cladinose sugar during acidic work-up. Therefore, catalytic reduction, as the method of choice for the large scale preparation of 9-deoxo-9a-aza-9a-homoerythromycin (**27**) was further optimized to use Pt/C as a catalyst in aqueous solvent (methanol/water/2N acetic acid).

The alkylation of nitrogen of *N*-oxide derivative **28**, with alkyl halides and sub-sequent catalytic hydrogenolysis (Scheme 2.8) provided a set of compounds with somewhat expanded antibacterial activities but also greatly improved pharmacoki-netic properties in comparison to erythromycin A [15]. Among them methylated derivative (9-deoxo-9a-methyl-9a-aza-9a-homoerythromycin A, named azithromy-cin) was found as the most promising candidate. The pharmacokinetic properties

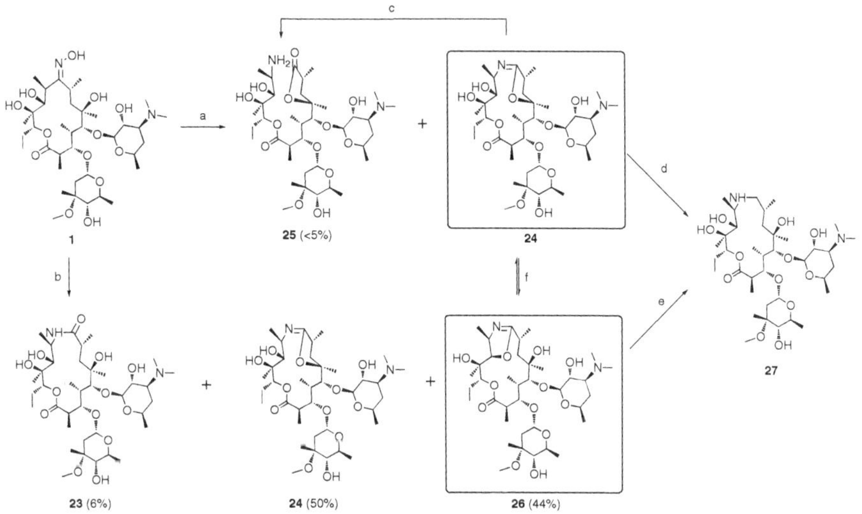

Scheme 2.7: The synthesis of an azalide scaffold. Reagents and conditions: a) TsCl, NaHCO$_3$, aq. acetone, 0 °C, 86% of **24**; b) TsCl, pyridine, ether, −45 °C, 85% of **24** and **26**; c) acetic acid/water; d) H$_2$, PtO$_2$, acetic acid, 80%, or NaBH$_4$, methanol, rt, acidic work-up, 60%; e) H$_2$, PtO$_2$, acetic acid, 86%; f) organic solvent.

of azithromycin are surprisingly good [16] as is its stability in acidic conditions [17]. An alternative methylation method, used in the industrial process of azithromycin production, is reductive *N*-methylation of 9-deoxo-9a-aza-9a-homoerythromycin **27** using Eschweiler-Clarke conditions (conditions: c, Scheme 2.8) [18].

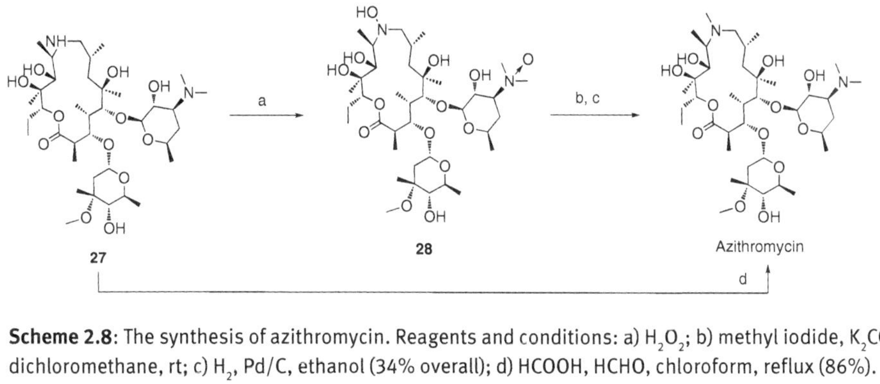

Scheme 2.8: The synthesis of azithromycin. Reagents and conditions: a) H$_2$O$_2$; b) methyl iodide, K$_2$CO$_3$, dichloromethane, rt; c) H$_2$, Pd/C, ethanol (34% overall); d) HCOOH, HCHO, chloroform, reflux (86%).

In an attempt to avoid the use of catalytic hydrogenation in the process of azithromycin preparation, the reduction of erythromycin A 6,9-imino ether **24** with metal hydrides was further investigated and developed as a useful alternative industrial method. It is based on the reduction of **24** with sodium borohydride at a low temperature

(–10 °C) that provides organoborate derivative **29** without the need for acidic work-up (Scheme 2.9) [19]. Such a derivative could be also successfully *N*-methylated under Eschweiler-Clarke conditions to provide **30** and then hydrolyzed to azithromycin.

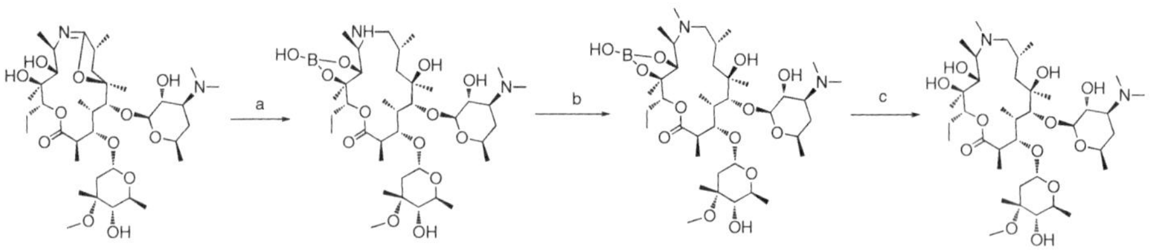

Scheme 2.9: An alternative synthesis of azithromycin. Reagents and conditions: a) NaBH$_4$, methanol, 0–10 °C, 92%; b) HCOOH, HCHO, chloroform, reflux, 57%; c) pH 2, 15 °C, 70%.

An interesting phenomenon of erythromycin 9-*E*-oxime **1**, which is almost exclusively formed by the oximation of erythromycin A, is that it readily isomerizes in basic conditions to 9-*Z*-oxime **2** (*Z*:*E* = 4:1) (Scheme 2.10) [20]. Solid *Z*-isomer **2** is stable but in organic solvents slowly isomerizes back to *E*-izomer **1**. This phenomenon allowed the synthesis of the azithromycin isomer, 9-deoxo-8a-aza-8a-homoerythromycin A (**36**, Scheme 2.10) that showed *in vitro* antibacterial activity equal to azithromycin. In both aqueous and anhydrous Beckmann conditions used, 6,9-iminoether **32** was obtained as a minor reaction product. In anhydrous conditions, 9,12-iminoether **33** was major product but it is in equilibrium, which depends on the solvent used, with its 10-*epi* analogue **34**. However, in aqueous conditions lactam **31** is the major product. The reduction of both iminoethers **32** and **33** afforded 9-deoxo-8a-aza-8a-homoerythromycin A (**35**), which is methylated using Eschweiler-Clarke conditions to provide **36**.

2.3.4 Clarithromycin/azithromycin hybrids

The superior properties of clarithromycin and azithromycin over parent erythromycin A led to an idea of the combination of both in one molecule. Obviously, the fastest synthesis of such a molecule should involve direct methylation of the 6-OH moiety of azithromycin [21]. However, it turned out that the methylation of 2',4"-protected azithromycin **37** (Scheme 2.11) with methyl iodide in the presence of a base does not follow the same pattern as the previously described methylation of erythromycin analogue **3** (see Scheme 2.2). Although it was initially claimed that the desired 6-*O*-methyl derivative **38** is the major product among variously methylated reaction products, it was later found that **38** is not formed at all and the actual main product of this reaction is 12-*O*-methyl derivative **39** [22]. Moreover, 6-OH was found to be the least reactive hydroxyl group of 2'-*O*-,3'-*N*-bis(benzyloxycarbonyl)-*N*-demethyl azithromycin **37**.

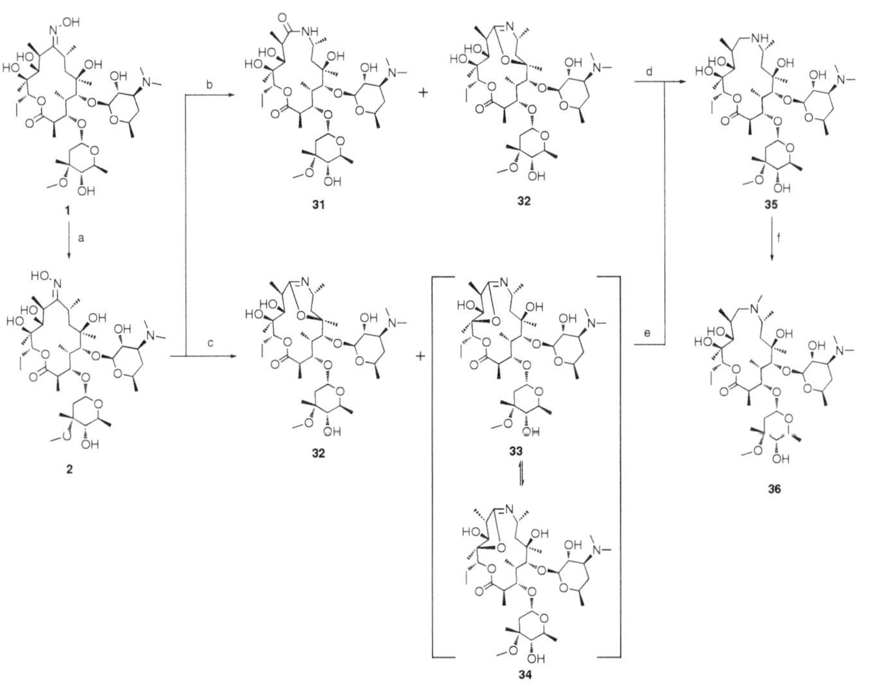

Scheme 2.10: The synthesis of 9-deoxo-8a-aza-8a-homoerythromycin A. Reagents and conditions:
a) LiOHxH$_2$O, ethanol (74%); b) TsCl, NaHCO$_3$, aq. acetone (48% **31**, 17% of **32**); c) TsCl, pyridine,
ether, 25% of **32** and 70% of mixture **33** and **34**; d) H$_2$, PtO$_2$, acetic acid, 42%; e) NaBH$_4$, ethylene
glycol, 45%; f) HCOOH, HCHO, chloroform, reflux, 85%.

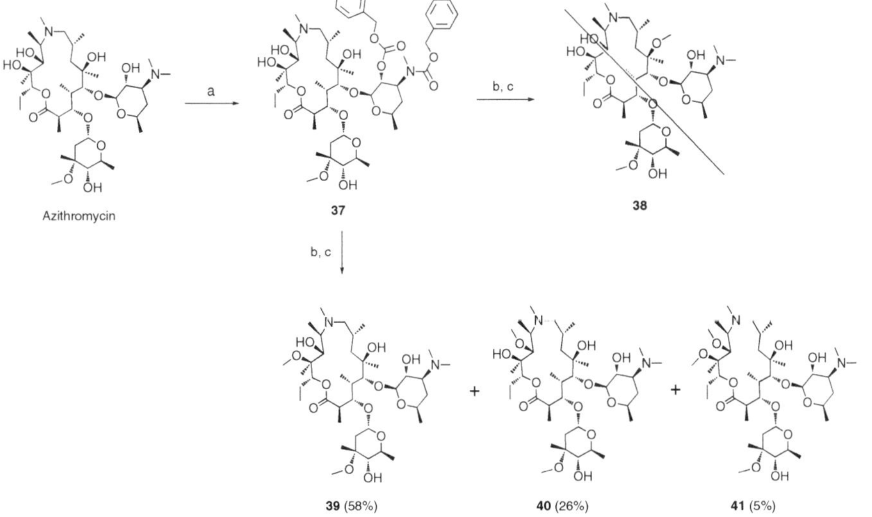

Scheme 2.11: The direct alkylation of 2'-*O*-,3'-*N*-bis(benzyloxycarbonyl)-*N*-demethyl azithromycin. Reagents
and conditions: a) benzyl chloroformate, NaHCO$_3$, benzene, 78%; b) CH$_3$I, NaH, DMSO/THF (1:1), 0–5 °C.

As all attempts to directly alkylate the 6-OH of azithromycin failed, the desired hybrid compound was prepared starting from clarithromycin (Scheme 2.12) [23, 24]. Interestingly, the stereospecific formation of 9-*E*-oxime **42** required elevated temperatures and an extended reaction time. In contrast to erythromycin 9-*E*-oxime **1**, clarithromycin 9-*E*-oxime **42** does not readily isomerize to 9-*Z*-oxime. As the formation of the 6,9-imino ether derivative is blocked in **42**, the Beckmann rearrangement to desired azalide **44** relied on the formation of 9,11-imino ether **43**. Again, contrary to the described conversion of erythromycin 9-*E*-oxime **1** to 6,9- and 9,11-imino ethers **24** and **26** (see Scheme 2.7), the standard aqueous Beckmann rearrangement of **42** produced a solely macrolactam **46**, while an anhydrous reaction in pyridine provided the desired 9,11-imino ether **43**. The subsequent reduction of imine **43** and Eschweiler-Clarke methylation of **44** provided the desired 6-*O*-methyl azithromycin **45**. Unfortunately, the antibacterial activity of this compound was weaker in comparison to azithromycin and clarithromycin.

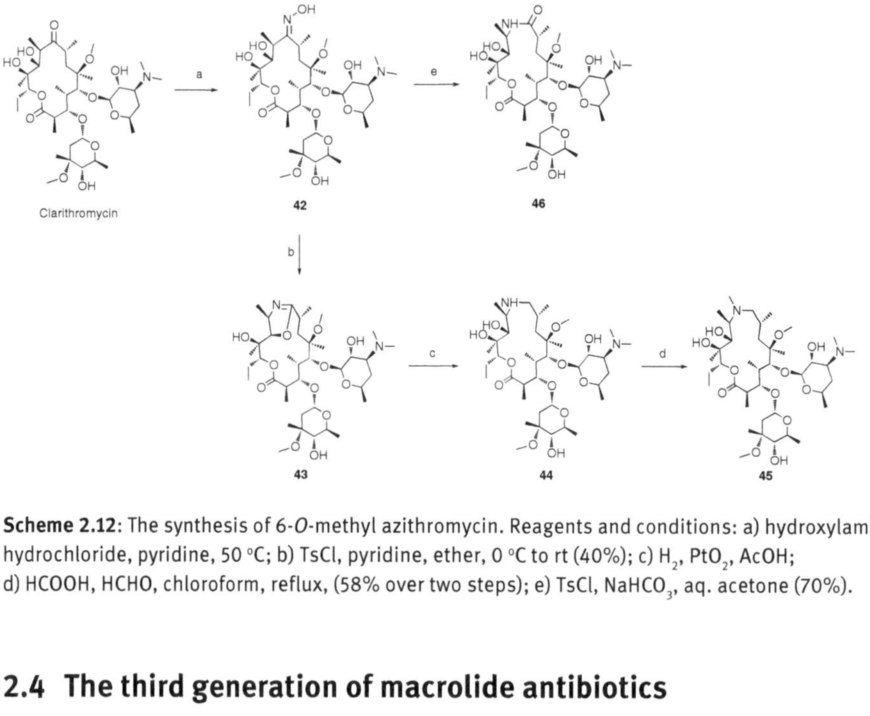

Scheme 2.12: The synthesis of 6-*O*-methyl azithromycin. Reagents and conditions: a) hydroxylamine hydrochloride, pyridine, 50 °C; b) TsCl, pyridine, ether, 0 °C to rt (40%); c) H$_2$, PtO$_2$, AcOH; d) HCOOH, HCHO, chloroform, reflux, (58% over two steps); e) TsCl, NaHCO$_3$, aq. acetone (70%).

2.4 The third generation of macrolide antibiotics

An incentive for the third generation of semisynthetic macrolide antibiotics was primarily set up to find novel antibiotics with broader antibacterial spectra, including higher efficacy against resistant bacteria, especially resistant *Streptococcus pneumoniae* strains. To overcome the weak binding of "older" macrolides to base-specifically methylated or dimethylated bacterial ribosomes, which is the most widespread mechanism of macrolide resistance in bacteria, some more prominent structural changes of the initial erythromycin macrolide scaffold were needed. The main hallmark of the third generation of macrolide antibiotics is the lack of cladinose sugar and the formation of a new keto group at the C3 position of the erythromycin scaffold. Such compounds, named ketolides, were already isolated from natural sources, even before the erythromycins, but their antibacterial activity was quite weak. However, the fact that natural ketolides, in contrast to erythromycins, do not induce macrolide resistance [25] inspired the research around the 3-keto derivatives of erythromycin A.

Although the synthesis of erythromycin ketolide **49** *via* oxidation of the 3-OH of decladinosyl derivative **48** was reported in the era of erythromycin A studies [26], 2 decades later it was found that instead of 3-keto derivative **49**, 3,6-hemiacetal **50**, which is not active against bacteria, was formed (Scheme 2.13).

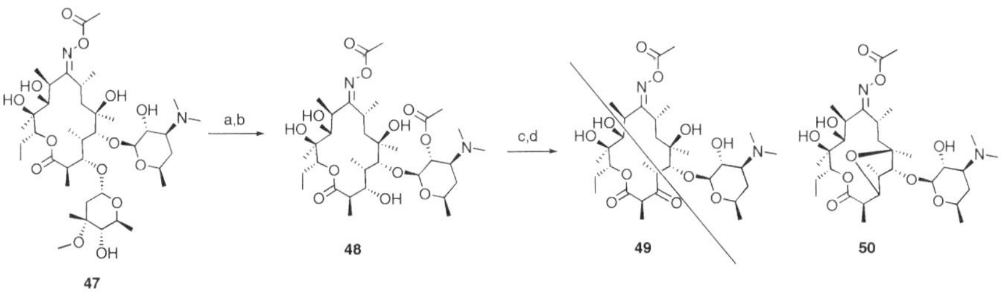

Scheme 2.13: Attempted synthesis of erythromycin ketolide. Reagents and conditions: a) HCl/H$_2$O; b) Ac$_2$O, K$_2$CO$_3$, acetone; c) CrO$_3$, H$_2$SO$_4$, H$_2$O, acetone; d) methanol (22% overall).

As it turned out that ketolides could not be formed on an erythromycin A scaffold having an unprotected C6-hydroxy group, the solution was found in the formation of ketolide moiety starting from 6-*O*-methylated macrolides (Scheme 2.14), primarily clarithromycin and 6-*O*-methyl roxithromycin (**51**) as well as 6-*O*-methyl azithromycin (**44**) [24, 27]. Acid-promoted cleavage of cladinose sugar provided decladinosyl derivatives **52**–**54**. Subsequent protection of 2'-OH, followed by the oxidation of 3-OH provided the desired ketolides **58**–**60** in good yields. Interestingly, azithromycin ketolide **60** was essentially antibacterially inactive, while clarithromycin ketolide **58** and 6-*O*-methyl roxithromycin ketolide **59** showed good antibacterial activity against susceptible strains but also some rather weak activity against erythromycin-resistant strains [27]. These promising results led researchers toward the synthesis of novel ketolides having enhanced activity against macrolide-resistant bacterial strains.

Another common feature of ketolides is the presence of 11,12-cyclic carbamate moiety annulated to the macrolactone ring. The first such macrolactone ring modification was reported on a clarithromycin molecule [28]. After protection of the 2'-OH and 4"-OH groups (**61**), introduction of the 11,12-cyclic carbamate moiety was accomplished using the two-step strategy (Scheme 2.15). The formation of intermediate 11,12-cyclic carbonate **62** was followed by β-elimination to afford 10,11-anhydro derivative **63**, which was trapped with an excess of carbonyldiimidazole to afford key 12-acylimidazolyl intermediate **64**. The base-induced intramolecular Michael

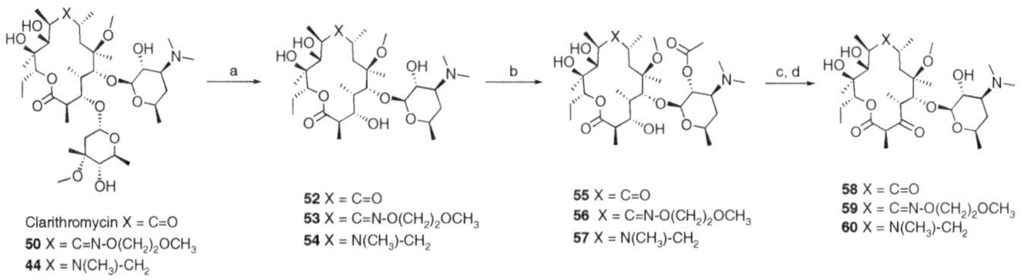

Clarithromycin X = C=O
50 X = C=N-O(CH₂)₂OCH₃
44 X = N(CH₃)-CH₂

52 X = C=O
53 X = C=N-O(CH₂)₂OCH₃
54 X = N(CH₃)-CH₂

55 X = C=O
56 X = C=N-O(CH₂)₂OCH₃
57 X = N(CH₃)-CH₂

58 X = C=O
59 X = C=N-O(CH₂)₂OCH₃
60 X = N(CH₃)-CH₂

Scheme 2.14: The synthesis of clarithromycin, roxithromycin and azithromycin ketolides. Reagents and conditions: a) HCl/H₂O; b) Ac₂O, K₂CO₃, acetone; c) for **55** and **56**: EDAC, HCl, DMSO, pyridinium trifluoroacetate, DCM, for **57**: CrO₃, H₂SO₄, H₂O, acetone; d) methanol (44% for **58**, 53% for **59**, 41% for **60** overall).

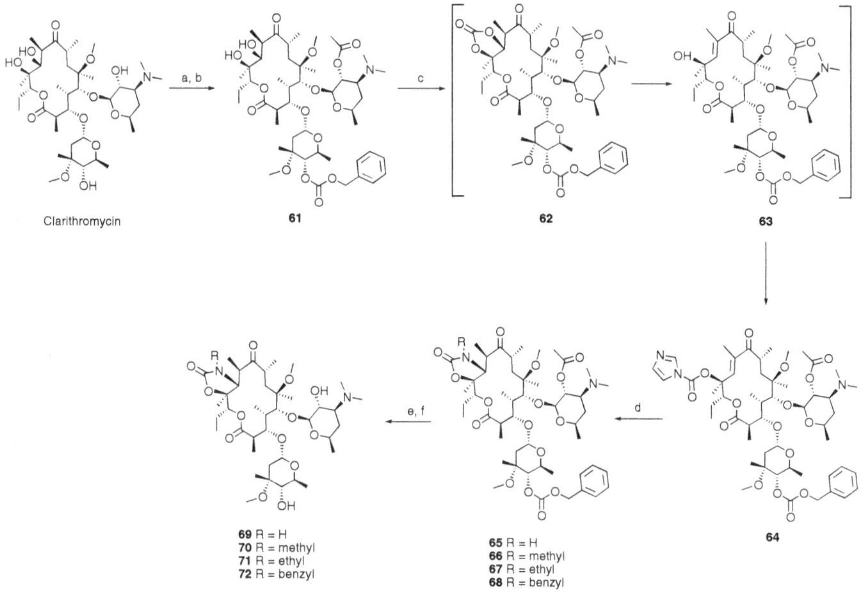

Clarithromycin

61

62

63

69 R = H
70 R = methyl
71 R = ethyl
72 R = benzyl

65 R = H
66 R = methyl
67 R = ethyl
68 R = benzyl

64

Scheme 2.15: The synthesis of clarithromycin 10,11-cyclic carbamates. Reagents and conditions: a) Ac₂O, triethylamine; b) CbzCl, DMAP, −20 °C; c) NaHMDS, −40 °C, 1,1'-carbonyldiimidazole, rt; d) for **65** tBuOK, for **66-68** R-NH₂, aq. acetonitrile, rt; e) methanol, rt; f) H₂, Pd/C.

addition to form 11,12-cyclic carbamate **65** was found to be stereoselective and afforded only one (natural) isomer at the C10 position. Subsequent deprotection of 2'-OH and 4"-OH afforded clarithromycin 11,12-cyclic carbamate **69**. Importantly, when the key 12-acylimidazolyl intermediate **64** was reacted with different amines in aqueous acetonitrile, stereoselective intramolecular cyclization also occurred to afford a small library of 10,11-cyclic carbamates bearing different substituents at the nitrogen. The clarithromycin 11,12-carbamate analog **72**, bearing the phenyl ring on the side chain, also showed moderate activity against the constitutively and inducibly resistant strains relative to clarithromycin [29].

2.4.1 Telithromycin (HMR3647)

The first approved ketolide, telithromycin, could be imagined as a hybrid between clarithromycin ketolide and clarithromycin 11,12-cyclic carbamate [27, 30]. The synthesis was started by the protection of 2'-OH of clarithromycin ketolide **58** (Scheme 2.16). However, the reaction sequence that worked well for the synthesis of 12-acylimidazolyl clarithromycin **64** (see Scheme 2.15) provided an unacceptably low yield of 12-acylimidazolyl clarithromycin ketolide analogue **75**. Therefore, the synthesis of key intermediate **75** was achieved in a three-step sequence *via* base-induced β-elimination of the mesylate formed at position C11 (**73**) in an overall yield of around 50% (Scheme 2.16). Further reaction of 12-acylimidazolyl intermediate **75** with ammonia afforded a mixture of natural and non-natural isomers **76** and **77** in a ratio of 2:1, respectively, while reactions with bulkier amino compounds provided stereospecifically only natural isomers. Therefore, the reaction of 12-acylimidazolyl ketolide **75** with 4-(3-pyridinyl)-1*H*-imidazole-1-butanamine (**78**) afforded only natural isomer named telithromycin in a 50% yield. Telithromycin showed very good activity against macrolide-resistant bacterial strains as well as good pharmacokinetic behavior.

A modified Pfitzner-Moffat procedure was used for selective 3-OH oxidation of decladinosyl derivative **55** (see Scheme 2.14) in the presence of free C11- and C12-hydroxy groups. However, this process uses a large quantity of 1-ethyl-3-(3-dimethylaminopropyl)carbodiimide hydrochloride (EDAC), which is an expensive reagent and difficult to obtain on a commercial scale. It was also found that this oxidation procedure provides variable yields and 9,12-hemiacetal as a major side product that could not be easily separated from desired 3-keto analogue **58** [31]. To overcome the described problems, the protection of both 11-OH and 12-OH of decladinosyl derivative **55** in the form of cyclic carbonate before the oxidation process was envisioned (Scheme 2.17). As only 3-OH of decladinosyl derivative **79** is left unprotected, most of the usual oxidation reagents could be used for conversion of **79** to ketolide **80** without any by-products. Moreover, the 11,12-carbonate group is a moderate leaving group and treatment with 1,8-diazabicyclo[5.4.0]undec-7-ene (DBU) as a base provides the desired 10,11-anhydro ketolide **74** in an overall yield of 74%.

More recently, a new procedure for the preparation of telithromycin in 33% yield from clarithromycin was reported (Scheme 2.18) [32]. The key change was the introduction of cyclic carbonate to position 11,12 (derivative **82**) before cleavage of the cladinose sugar. The oxidation of 3-OH of **79** was achieved in mild conditions using DMSO activated with P_2O_5. Cyclic carbonate protection was then eliminated using sodium bis(trimethylsilyl)amide (NaHMDS) as a base to provide 10,11-anhydro ketolide **74** that reacts with an excess of base and CDI to afford 12-*O*-acyl imidazolide **75**. Telithromycin was then prepared in the already described manner.

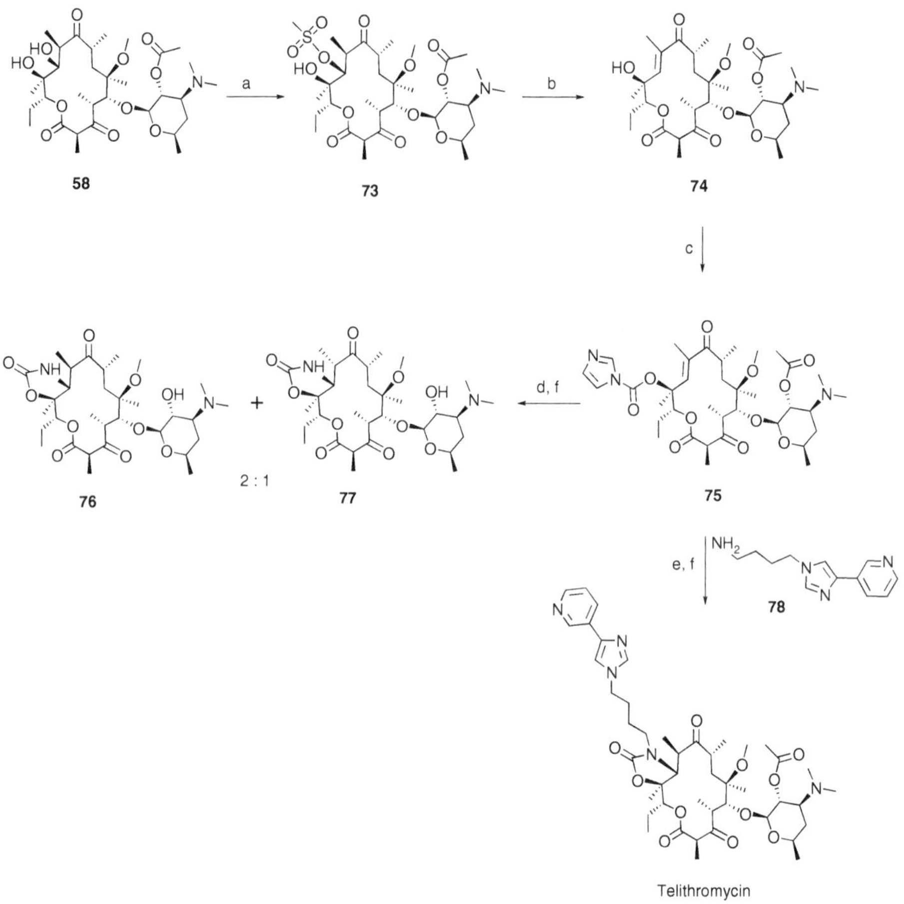

Scheme 2.16: The synthesis of telithromycin. Reagents and conditions: a) mesyl anhydride, pyridine (79%); b) DBU, acetone, r.t. (88%); c) NaH, 1,1'-carbonyldiimidazole, DMF, −10 °C (67%); d) NH$_3$, water, −40 to 20 °C; e) **78**, acetonitrile/water, 60 °C; f) methanol (50% over two steps).

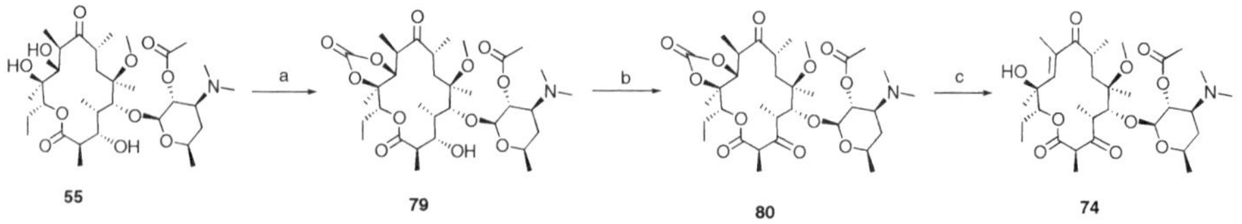

Scheme 2.17: Improved synthesis of intermediate 10,11-anhydro ketolide **74**. Reagents and conditions: a) bis(trichloromethanol)carbonate ("Triphosgene"), pyridine, DCM (95%); b) for example: N-chlorosuccinimide, dimethylsulfide, trimethylamine, DCM, −5 °C (94%) or piridinium chlorochromate, DCM, 30 °C, (96%); c) DBU, acetone, reflux (81%).

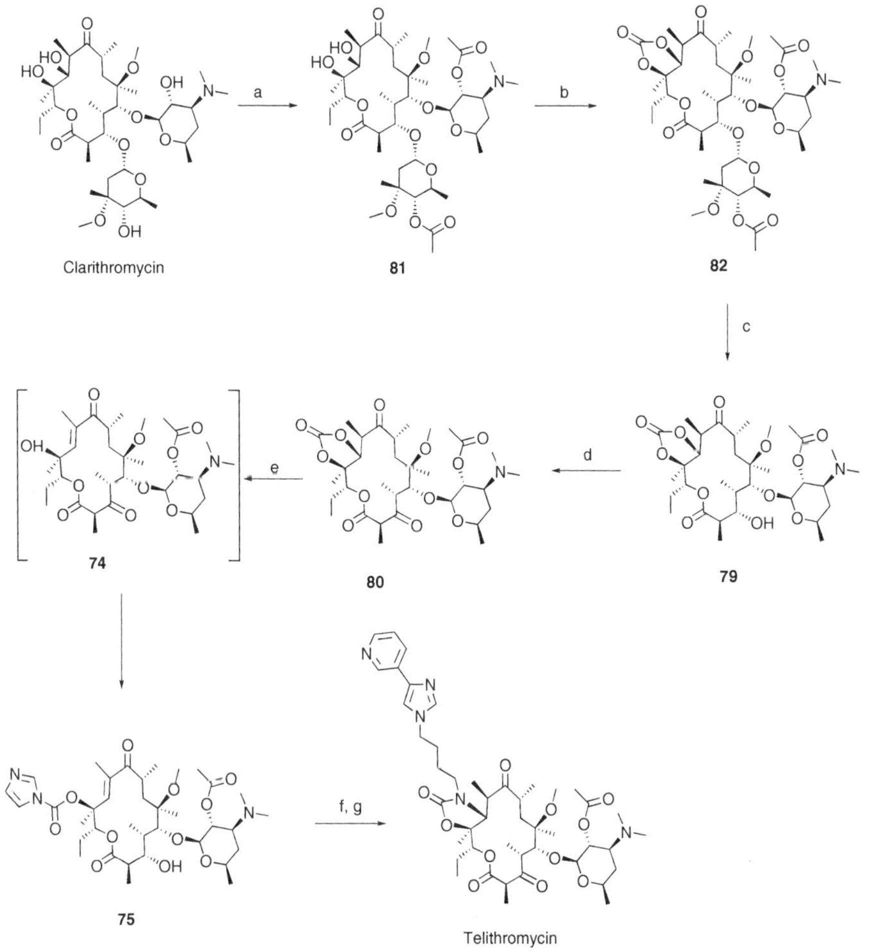

Scheme 2.18: Novel synthesis of telithomycin. Reagents and conditions: a) acetic anhydride, DMAP, DCM (91%); b) bis(trichloromethanol)carbonate ("Triphosgene"), pyridine, DCM (89%); c) 1N HCl, ethanol, rt (90%); d) DMSO, P$_2$O$_5$, trimethylamine, DCM (83%); e) NaHMDS, 1,1'-carbonyldiimidazole, THF, 0 °C to rt (75%), f) 4-(3-pyridinyl)-1H-imidazole-1-butanamine (**78**), acetonitrile/water, 50 °C; f) methanol (72% over two steps).

2.4.2 Cethromycin (ABT773)

Conformational studies of telithromycin and other macrolides revealed that if the aryl-alkyl group is tethered to a C6 position, instead to a 11,12-cyclic carbamate, it would adopt a very similar spatial position on the hydrophilic face of the macrol-actone ring [33]. As direct alkylation of sterically hindered 6-OH with larger alkyl groups is quite tedious, the synthesis of such ketolides started with the introduction of 6-O-allyl moiety onto an erythromycin scaffold (Scheme 2.19). An allyl group can be further functionalized with different alkyl-aryl moieties using Heck coupling conditions. The desired 6-O-allyl macrolide **84** was prepared by direct alkylation for the synthesis of 6-O-alkyl erythromycin derivatives (see Scheme 2.5) in a yield of 33% over three steps. The introduction of 11,12-cyclic carbamate could be performed prior to or after formation of 3-keto moiety. As the intramolecular Michael addition of 12-acylimidazolyl derivatives provides a mixture of C10 epimers on ketolide scaffold and mostly natural C10 isomer if cladinose is still present (compare Schemes 2.15 and 2.16), 6-O-allyl erythromycin A (**84**) was firstly converted to 12-acylimidazolyl derivative **86**. The intramolecular Michael addition then provided good selectivity towards formation of the desired natural C10 isomer **87** (<10% of C10-epimer **88** was formed). The subsequent sequence for the ketolide formation provided 6-O-allyl-11,12-cyclic carbamate-3-keto derivative **89** in 80% yield over two steps. Pd-catalyzed Heck coupling with aryl halogenides was used for the introduction of different aryl-alkyl groups to provide solely *trans*-6-O-aryl derivatives. A 3-Quinolyl aryl derivative, named cethromycin, demonstrated improved *in vitro* and *in vivo* antibacterial activity against resistant bacteria when compared to telithromycin.

Cethromycin was prepared using the above route in an overall yield of 11%–16% starting from erythromycin. The allylation of 6-OH and the Heck coupling seemed to be major obstacles in the synthetic sequence. Alkylation with allyl bromide, as described above, suffers from variable yields due to significant formation of diallylated products formed after partial cleavage of trimethylsilyl protections [34]. An alternative procedure using Pd-catalyzed allylation with allyl *t*-butylcarbonate afforded a much better yield (77%) of 6-O-allyl derivative **83** (Scheme 2.20). Moreover, it was found that direct Pd-catalyzed allylation of **17** with propenyl quinolyl *t*-butylcarbonate (**90**) proceeds in excellent yield (92%) to afford the directly desired substituent at the C6 position without the formation of any regio- or geometrical isomers.

Pd-catalyzed allylation of **17** with propenyl quinolyl *t*-butylcarbonate (**90**) improved the yield of the synthesis of cethromycin but for successful large scale production, the protection/deprotection strategy needed further optimization. The goal was to improve stability in the acidic conditions required for deoximation and to avoid reprotection of 2'-OH and 4"-OH (sequence **83** to **85**, Scheme 2.19). Therefore, instead of silyl protections, a tribenzoate protecting scheme was employed (Scheme 2.21) [35]. The advantages of benzoate protection are the ease of preparation and the tendency to crystallize, which expedites isolation and purification of intermediates. The oxime hydroxyl of tribenzoate intermediate **93** could be selectively deprotected to afford **94**. Deoximation of **94** in optimized conditions improved the yield of ketone **95** to 76%. Improved cethromycin synthesis also employs a modified method for the introduction of 11,12-cyclic carbamate moiety. This was achieved in efficient one-pot three-reaction manner to provide variable mixtures of C-10 epimeric 11,12-cyclic carbamates that equilibrated under reaction conditions to a 98:2 mixture in favor of natural

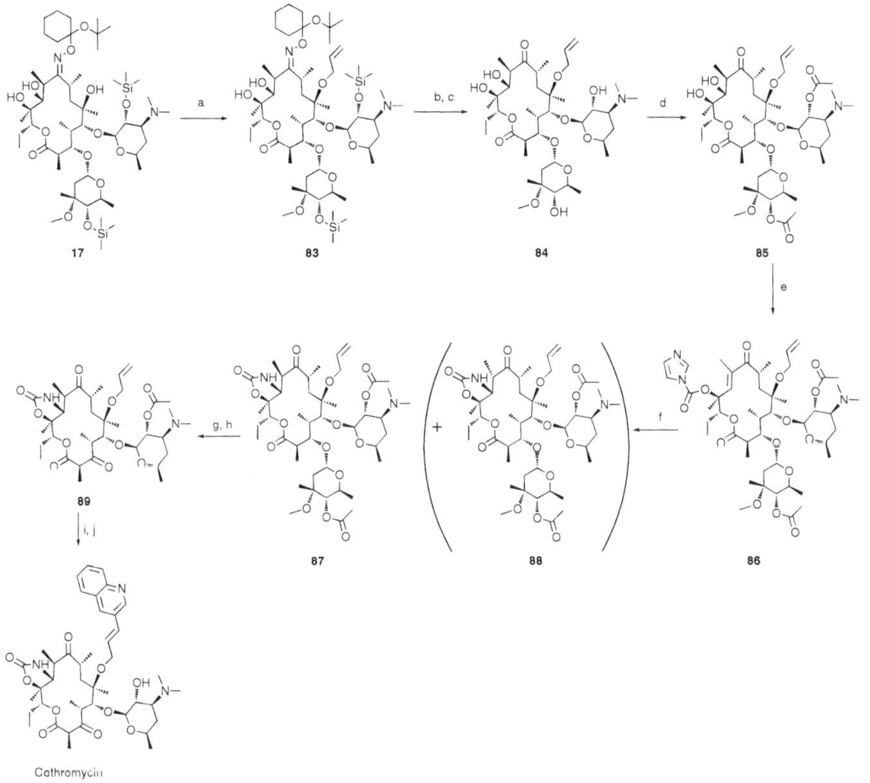

Scheme 2.19: The first synthesis of cethromycin. Reagents and conditions: a) allyl bromide, tBuOK, DMSO/THF; b) AcOH, aq. acetonitrile; c) NaHSO$_3$/HCOOH, ethanol/water (33% over three steps); d) acetic anhydride, trimethylamine, DMAP, DCM; e) NaHMDS, 1,1'-carbonyldiimidazole, THF/DMF; f) aq. NH$_3$, acetonitrile/THF (78% over three steps); g) HCl, ethanol; h) N-chlorosuccinimide, dimethylsulfide, trimethylamine, DCM (80% over two steps); i) 3-quinolyl bromide, Pd(OAc)$_2$, P(o-tolyl)$_3$, triethylamine, acetonitrile; j) methanol (60%–85% over two steps).

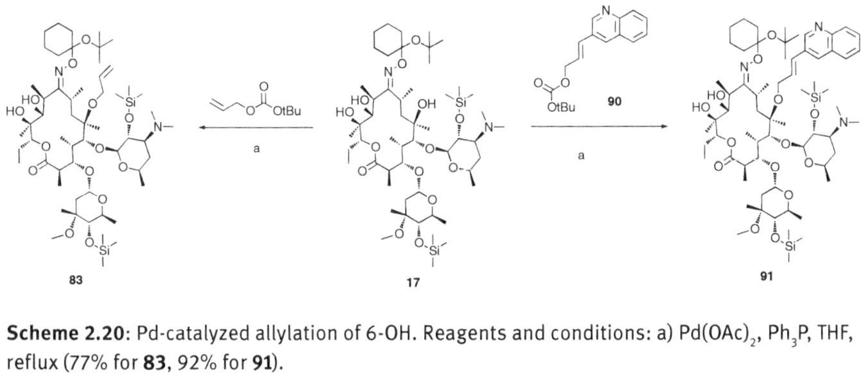

Scheme 2.20: Pd-catalyzed allylation of 6-OH. Reagents and conditions: a) Pd(OAc)$_2$, Ph$_3$P, THF, reflux (77% for **83**, 92% for **91**).

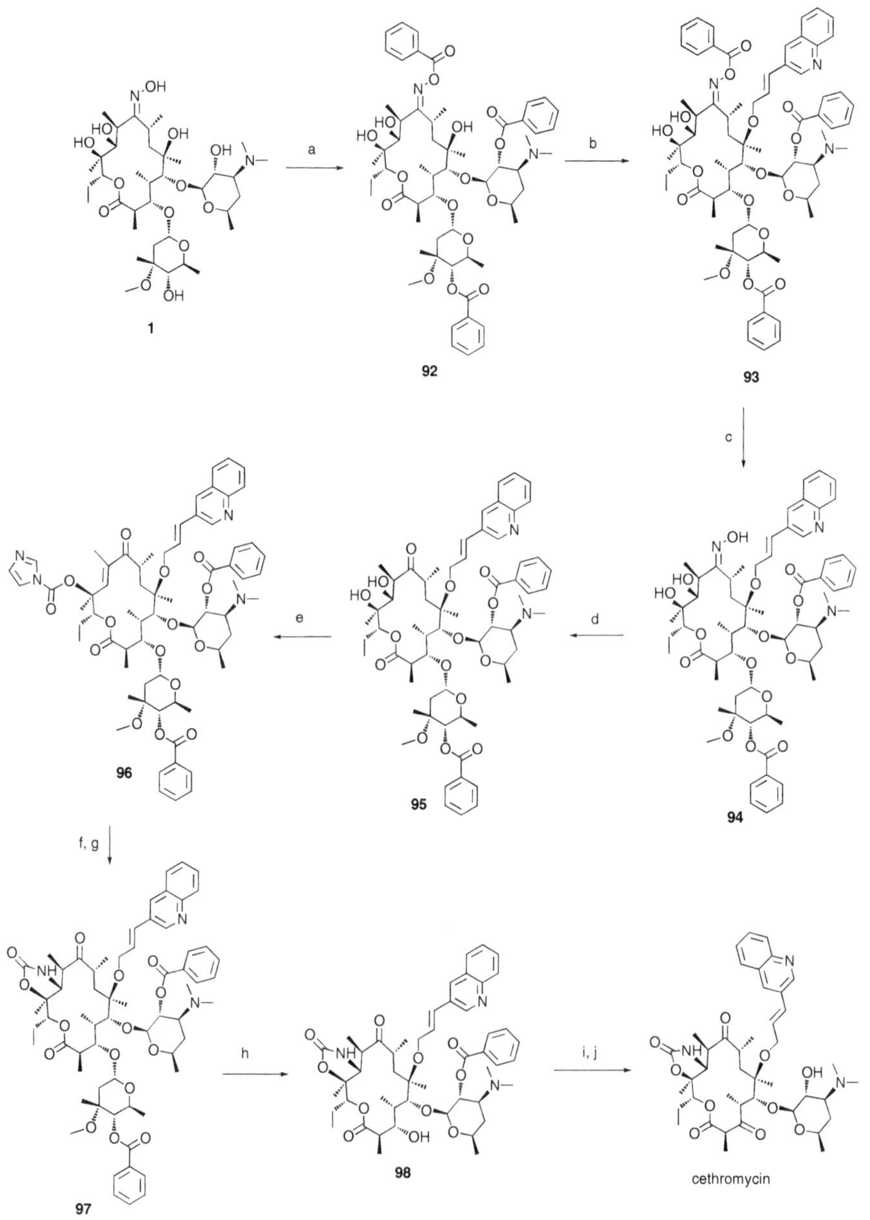

Scheme 2.21: Improved synthesis of cethromycin. Reagents and conditions: a) benzoic anhydride, trimethylamine, DMAP, THF (82%); b) [3-(3-quinolyl)-2-propenyl-1-*tert*-butyl carbonate, Pd$_2$(dba)$_3$, dppb, THF; c) 1 M NaOH, isopropanol, 0 °C (94% over two steps); d) L-tartaric acid, NaHSO$_3$, THF/water, 90 °C (76%); e) NaHMDS, 1,1'-carbonyldiimidazole, THF/DMF; f) NH$_3$; g) tBuOK (87% over three steps); h) 2 N HCl, ethanol (100%); i) *N*-chlorosuccinimide, dimethylsulfide, trimethylamine, DCM (97%); j) methanol, reflux (92%).

stereoisomer **97**, which was then easily isolated by crystallization in a yield of 87%. Various oxidation methods of 3-OH were tried but the Corey-Kim method proved the best, although due to the water-sensitivity of the Corey-Kim reagent, efficient drying of decladinosyl macrolide **98** was critical. The overall yield of cethromycin was thus improved to 40% starting from erythromycin A 9-*E*-oxime (**1**).

2.4.3 Solithromycin (CEM101)

Solithromycin is the first fluoroketolide member that has gone through clinical studies [36]. It is quite similar to telithromycin, having an aryl-alkyl chain tethered to 11,12-cyclic carbamate. Instead of pyridine connected to an alkyl chain *via* imidazole, solithromycin contains aminophenol and triazole moieties. A distinct feature of solithromycin is the presence of fluorine at position C2 that prevents the 3-keto enolization observed in other ketolides [37]. The patented synthesis of solithromycin starts again from clarithromycin in a similar fashion to the synthesis of telithromycin (Scheme 2.22) [38]. The introduction of the 11,12-cyclic carbamate bearing alkylazide side chain was done on dibenzoate protected intermediate **99** prior to cladinose cleavage to provide **101**. Subsequent 3-keto generation *via* Corey-Kim or Dess-Martin oxidation of decladinosyl derivative **102** provided ketolide **103**. Fluorine was then introduced to the C2 position using *N*-fluorobenzenesulfonimide as a mild electrophilic fluorinating reagent. Fluorination affords only *S*-isomer at the C2 position while the *R*-isomer was not observed and could not be prepared by a number of different approaches [39]. The copper(I)-catalyzed azide-alkyne cycloaddition reaction of azide moiety of **104** and terminal alkyne of **105** provided ketolide **106**. Subsequent methanolysis to deprotect 2'-OH afforded solithromycin.

2.4.4 Other ketolides: TE-802, modithromycin (EDP-420)

An interesting member of the ketolides with improved antibacterial activity against resistant bacterial strains is the tricyclic macrolide TE-802 (Scheme 2.23) [40]. The reaction of 12-acylimidazolide **107** with a large excess of ethylenediamine provides aminoethyl-substituted 11,12-cyclic carbamate **108**, predominantly having a natural *R*-configuration at position C10 (ratio of 10*R*:10*S* = 92:8). Interestingly, when the excess of ethylenediamine is decreased, stereoselectivity also decreases and could even be reversed (ratio of 10*R*:10*S* = 40:60) if only 2.5 eq of ethylenediamine is used. The formation of a diazaheptene ring *via* the 9,11 cyclization of **109** was performed in the presence of a small excess of formic or acetic acid to provide tricyclic macrolide **110** from which ketolide TE-802 was synthesized by the usual reaction sequence.

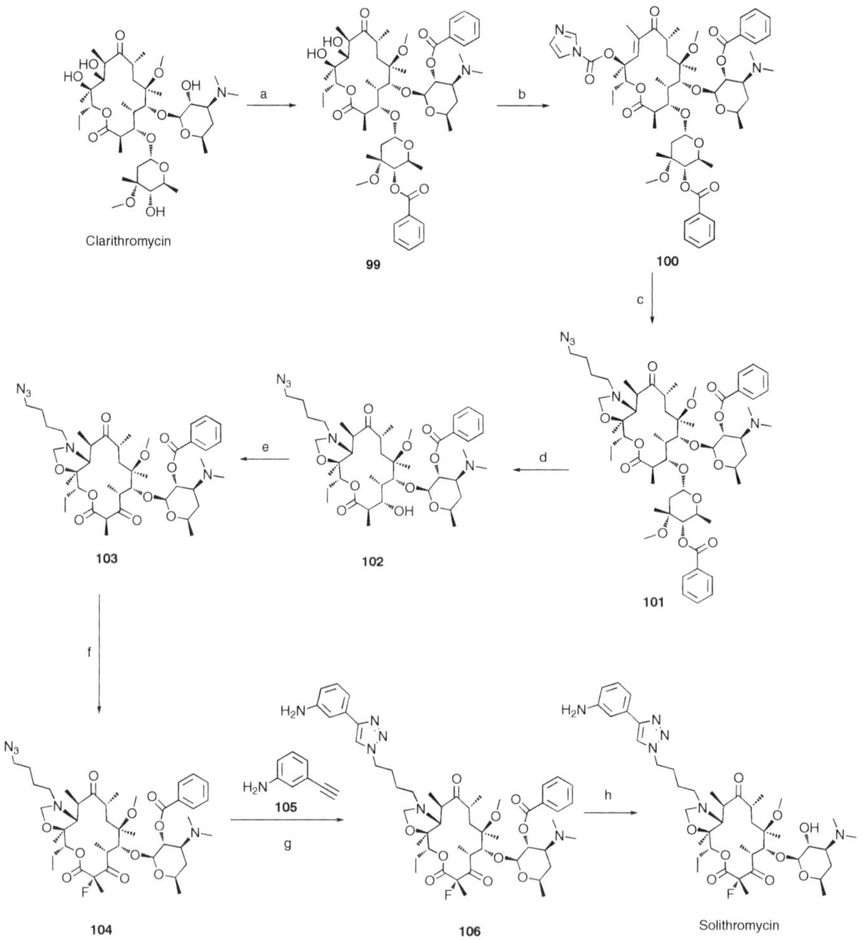

Scheme 2.22: The synthesis of solithromycin. Reagents and conditions: a) benzoic anhydride, trimethylamine, DMAP, THF; b) DBU, 1,1'-carbonyldiimidazole, DMF, rt; c) 4-azido butyl amine, DBU, DMF, 25–35 °C; d) HCl, acetone/water; e) N-chlorosuccinimide, dimethylsulfide, trimethylamine, DCM, 0 to −20 °C or Dess-Martin periodinane, DCM, 10–15 °C; f) N-fluorobenzenesulfonimide, tBuOK, THF, rt; g) **105**, CuI, diisopropylethylamine, acetonitrile, rt; h) methanol.

Another interesting example is the synthesis of the 6,11-bridged ketolide modithromycin (Scheme 2.24) [41]. The bridge between positions 6 and 11 prevents intramolecular cyclizations in acidic conditions but it is also a point for the attachment of the required aromatic side chains. The synthesis of modithromycin starts by the protection of three most reactive hydroxyl groups of erythromycin A 9-*E*-oxime. The bridge was introduced onto protected **114** *via* Pd-catalyzed tandem 6-*O* and 11-*O*-allylic dialkylation using bis(Boc)-protected allylic diol **115** [42]. After installation of the 6,11-bridge, cladinose and oxime protection were simultaneously hydrolyzed to provide **117** in 55%–65% overall yield. The yield is lower if the dryness and purity of the protected

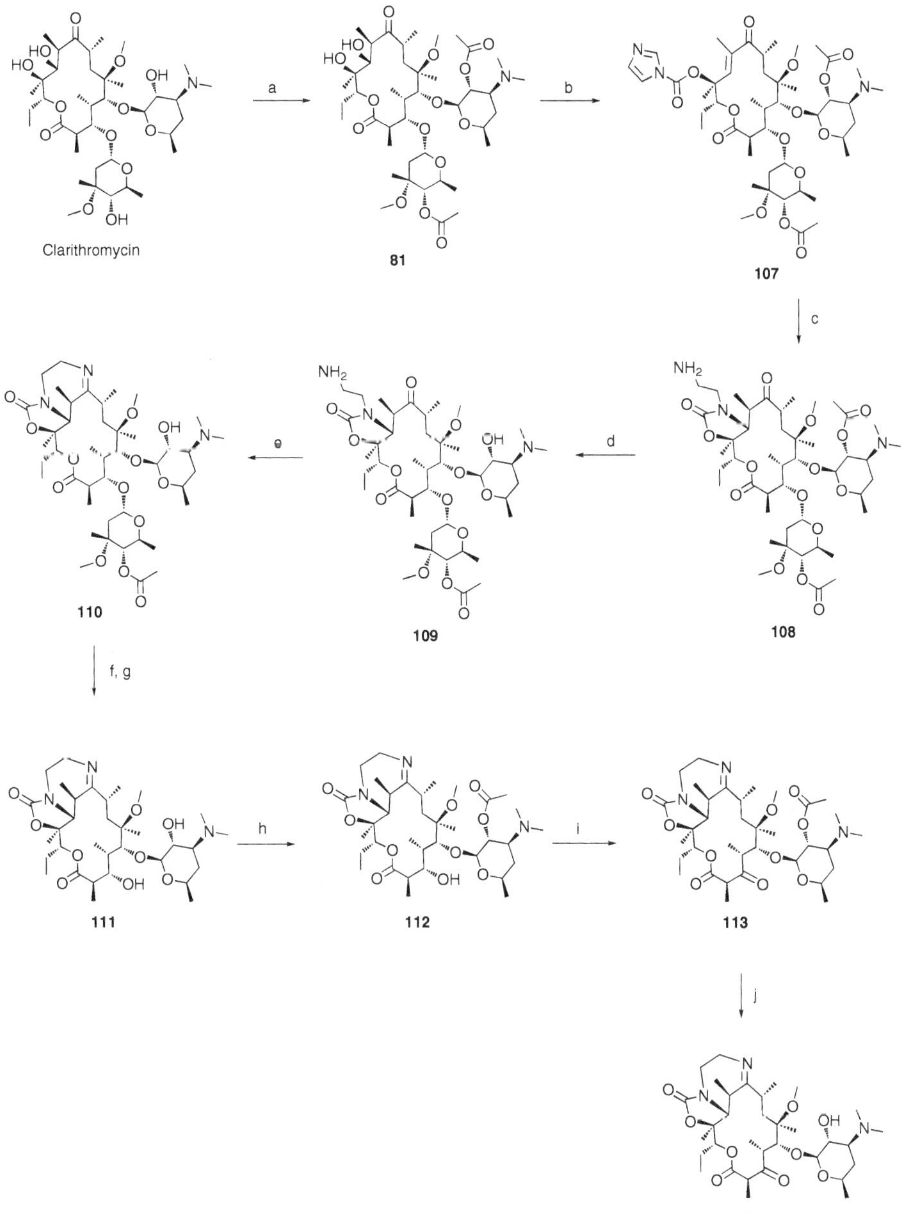

Scheme 2.23: The synthesis of TE-802. Reagents and conditions: a) acetic anhydride, DMAP, DCM (87%); b) NaH, 1,1'-carbonyldiimidazole, DMF/THF, rt (62%); c) ethylenediamine, acetonitrile, rt; d) methanol (91%, over two steps); e) acetic acid, ethanol, 60 °C (65%); f) DBU, methanol (48%); g) 2 N HCl, ethanol (89%); h) acetic anhydride, acetone (80%); i) Pfitzner-Moffat conditions: 1-ethyl-3-(3-dimethylaminopropyl)carbodiimide hydrochloride (EDAC x HCl), DMSO, pyridinum trifluoroacetate, DCM, rt; j) methanol (60% over two steps).

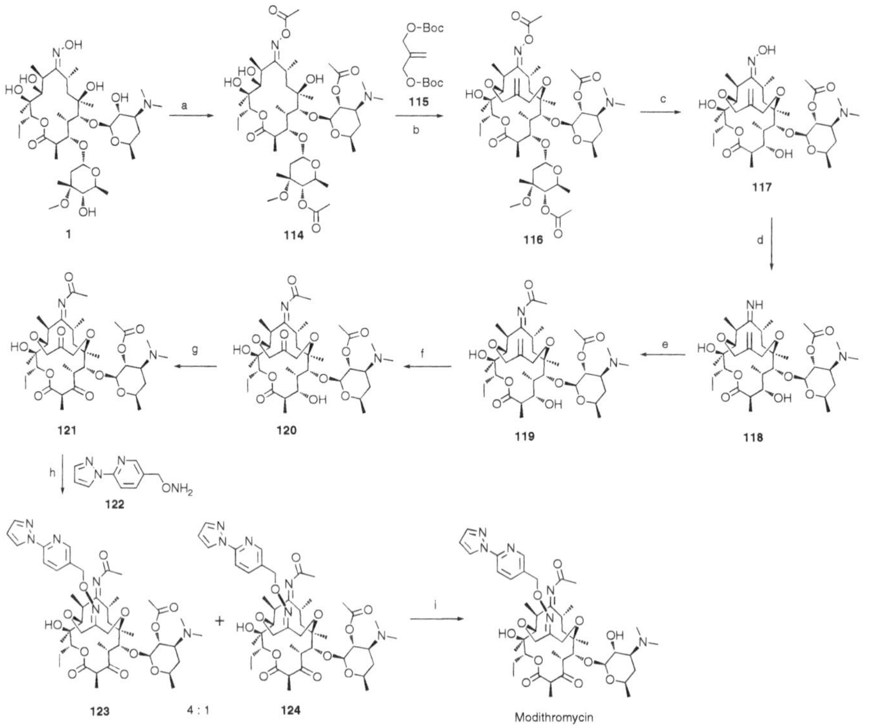

Scheme 2.24: Synthesis of modithromycin. Reagents and conditions: a) acetic anhydride, DMAP, TEA, ethyl acetate (80%); b) **115**, Pd$_2$(dba)$_3$, dppb, THF, reflux; c) 1 N HCl, THF, 60 °C, (50%–65% over two steps); d) TiCl$_3$ in aq. HCl, ethanol, rt; e) acetic anhydride, DCM, rt (84% over two steps); f) OsO$_4$, NaIO$_4$, acetone/water, 35–42 °C; g) N-chlorosuccinimide, dimethylsulfide, DCM, –20 °C (55% over two steps); h) **122**, 1 N HCl, ethanol/water, 0–5 °C; i) methanol (40% over two steps).

oxime **114** was not satisfactory before applying the Pd-catalyzed process. Obtained 9-E-oxime **117** is then reduced to 9-imine **118**, which is very stable and could not be further hydrolyzed to 9-keto analogue, and then acetylated to provide **119** in an 85%–90% yield. Subsequent two-step one-pot oxidations, firstly oxidative cleavage of the bridge double bond using the OsO$_4$-catalyzed NaIO$_4$ method and then the Corey-Kim oxidation of 3-OH, provided diketone **121** in a 55% yield. The reaction of diketone **121** with hydroxylamine **122** provides a mixture of E/Z oximes **123** and **124** (in a 4:1 ratio, respectively). After deprotection of 2'-OH by methanolysis, oxime isomers were separated by crystallization to provide E-isomer (modithromycin) in a 40%–42% yield.

2.5 Antibacterial macrolide hybrids – next generations of macrolide antibiotics

Activity against Gram-negative bacterial strains is still a weak point of macrolide antibiotics, including macrolides of both the second and third generations. The improvement of this weakness should be the target of the next macrolide generation. Another useful feature of novel macrolide generations would be improved activity against constitutively resistant bacterial strains. To fulfil these goals, new approaches to the (semi)synthesis of rationally designed macrolides are needed. The most recent such approach is synthesis of macrolide hybrids [43]. In general, hybrid molecules are formed by the incorporation of two pharmacophores into a single molecule that shows different biological properties in comparison to both pharmacophores alone. Recently, methods for the preparation of macrolide conjugate hybrids and macrolide merged hybrids have been described.

2.5.1 Macrolide conjugates

The conjugates of macrolide antibiotics with quinolyl moiety *via* linker to 4"-OH of cladinose belong to a novel class of antibacterial macrolide hybrids named macrolones. The first edition of macrolones contained a heteroalkyl linker connected to 4"-OH *via* ester moiety and shows excellent antibacterial activity against bacterial strains with *erm*-mediated resistance as well as against *Haemophilus infuenzae* [44, 45]. Following macrolones with 4"-*O*-alkyl, instead of 4"-*O*-acyl, linked quinolones also showed superior antibacterial activity [46]. An efficient and scalable process for the production of such macrolones was explored using two different synthetic routes, both relying on the key C-C coupling step, either *via* the Heck or Sonogashira method, and sharing the same intermediate **127** (Scheme 2.25). The 4"-γ-amino derivative **127** was prepared by alkylation of 4"-OH of 2'-*O*-acetylazithromycin 11,12-cyclic carbonate **125** with acrylonitrile and subsequent hydrogenation. Diazotation of **127** followed by allyloxylation of the intermediary diazonium salt provided intermediate **128** as a starting material for a Heck reaction with quinolones **129**. A reaction of the diazonium ion with propargyl alcohol provided intermediate **132** suitable for a Sonogashira reaction with quinolones **129** and also tricyclic quinolones **133**. Removal of 11,12-carbonate and 2'-*O*-acetyl groups was preferably done before performing coupling reactions. Hydrogenation of either Heck **130** or Sonogashira products **134** afforded the final macrolones **131** and **135**, respectively. The overall potency of the macrolones provides a clear advantage over currently used macrolide antibiotics, including telithromycin, which is inactive against constitutively resistant *Streptococcus pyogenes*.

2.5.2 Macrolide merged hybrids

A recently published synthetic platform uses simple building blocks and a convergent assembly process for the synthesis of rationally designed diverse macrolide scaffolds not limited by the structure of the natural product erythromycin A [47]. As an illustration of the usefulness of this platform, a route to the merged hybrid macrolides, specifically ketolides and azalides, is described (Scheme 2.26). It was expected that

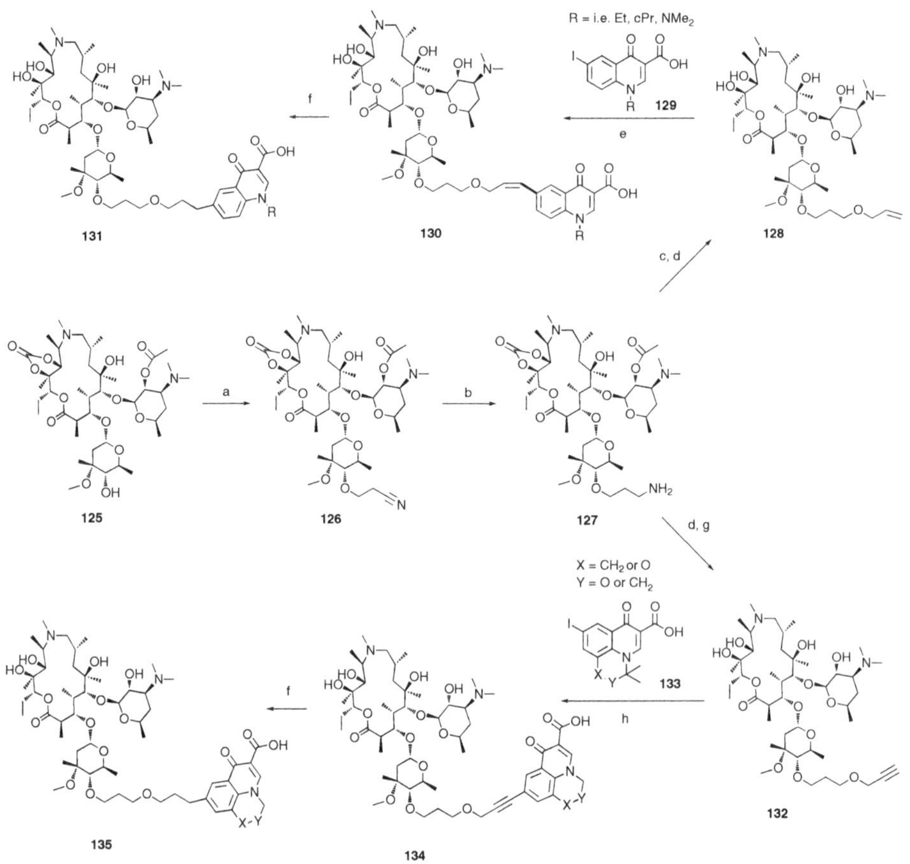

Scheme 2.25: The example of the synthesis of 4"-O-linked macrolones. Reagents and conditions:
a) acrylonitrile, tBuOH, NaH, 0 °C to rt, 24 h; b) acetic acid, 20 wt %, PtO$_2$, H$_2$, 5 bar, 18 h; c) allyl alcohol,
NaNO$_2$, HCOOH, 0 °C to rt, 24 h; d) methanol/water, K$_2$CO$_3$, 55 °C, 2 h; e) DMF, Pd(OAc)$_2$, P(o-tolyl)$_3$, **129**,
Et$_3$N, 65 °C, 2 h, 75 °C, 18 h; f) methanol, 10% Pd/C 10 wt %, H$_2$ 3 bar, 15 h; g) propargyl alcohol, HCOOH,
NaNO$_2$, 0–5 °C, 8 h; h) acetonitrile, CuI, triethylamine, **133**, Pd(PPh$_3$)$_2$Cl$_2$, 50 °C, 16 h.

the ketolide part would play a role in activity against resistant bacterial strains
while the azalide part could improve pharmacokinetics and also activity against
Gram-negative bacteria. It is worth noting that this approach allows the synthesis of
14-membered azalides that cannot be obtained by semisynthetic methods from erythromycin A. The convergent synthesis of such azaketolides starts with four simple
building blocks (**136–139**, Scheme 2.26). Key intermediates **140** and **141** were prepared
in seven steps in 57% and 58% yields, respectively, and joined by a reductive coupling
reaction affording an 82% yield of **142** in a diastereomeric ratio >20:1. Interestingly,
the key macrocyclization of intermediate **142** was very efficacious and generally provided azaketolide scaffold **143** in a 78% yield. An azide-alkyne dipolar cycloaddition
reaction provides the final azaketolide hybrid **144**.

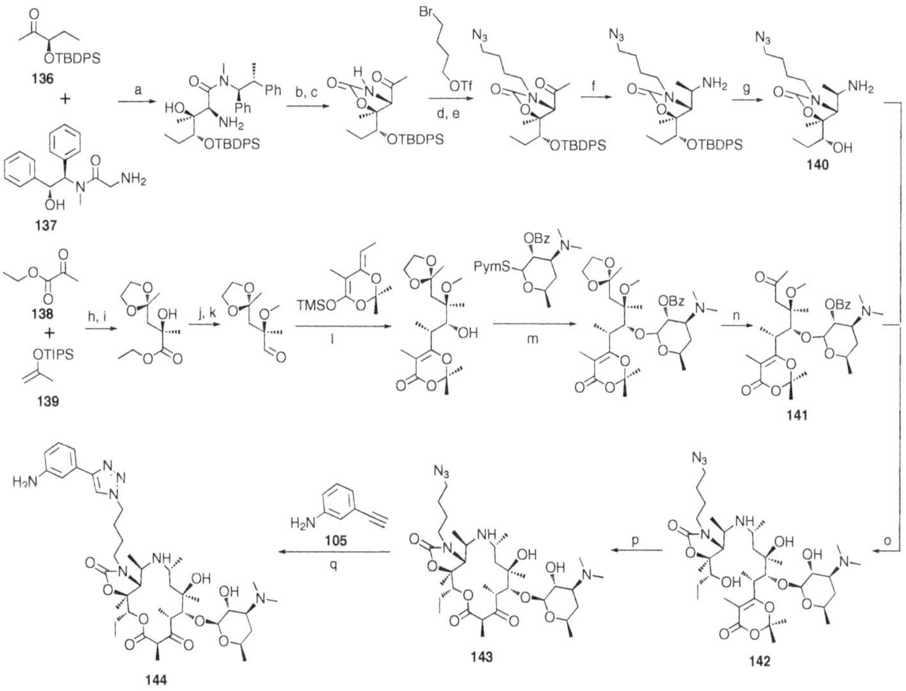

Scheme 2.26: The synthesis of 9-azaketolides. Reagents and conditions: a) LiHMDS, LiCl (98%); b) diisopropylethylamine, COCl$_2$; c) iPrMgCl, CH$_3$Li (76% over two steps); d) NaHMDS; e) NaN$_3$ (88% over two steps); f) NH$_3$, Ti(OiPr)$_4$, NaBH$_4$ (95%); g) Bu$_4$NF (92%); h) Pd[(S)-SegPhos]Cl$_2$, AgSbF$_6$ (93%, 92% e.e.); i) ethylene glycol, PPTS (92%); j) KH, MeI (97%); k) (iBu)$_2$AlH (96%); l) triethylamine, MgBr$_2$•OEt$_2$ (91%); m) AgOTf (81%, 16:1 β:α); n) HCl (100%); o) NaCNBH$_3$ (82%); p) 132°C, 1 mM, PhCl (78%); q) CuSO$_4$, sodium L-ascorbate (86%).

2.5.3 Macrolactone ring reconstruction

The ring reconstruction methodology of the macrolactone skeleton is also a very interesting platform that could enable the synthesis of novel classes of hybride macrolide antibiotics. An illustrative example is the synthesis of 11a-azalides possessing a variety of substituents in the C12 and/or C13 position [48–50]. The 11,12-diol moiety of the suitable protected erythromycin skeleton **146** can be oxidatively cleaved by treatment with lead tetraacetate to afford aldehyde **147** (Scheme 2.27). As β-ketoaldehyde moiety is unstable under reductive amination conditions, a subsequent reductive amination step with appropriate amino alcohols was conducted on structures bearing 9-OH or 9-E-oxime moiety instead to afford **148**. After successive hydrolysis of the remaining original C12–13 residue from **149**, the resulting acyclic skeleton **150** was cyclized by a macrolactonization reaction to provide 11a-azalide **151**. The structural diversity can be generated *via* insertion of appropriately functionalized amino alcohols or *via* N-alkylation of an inserted secondary amine. An interesting example is the preparation of 11a-azaketolide **155** bearing an aryl side chain attached to the C13.

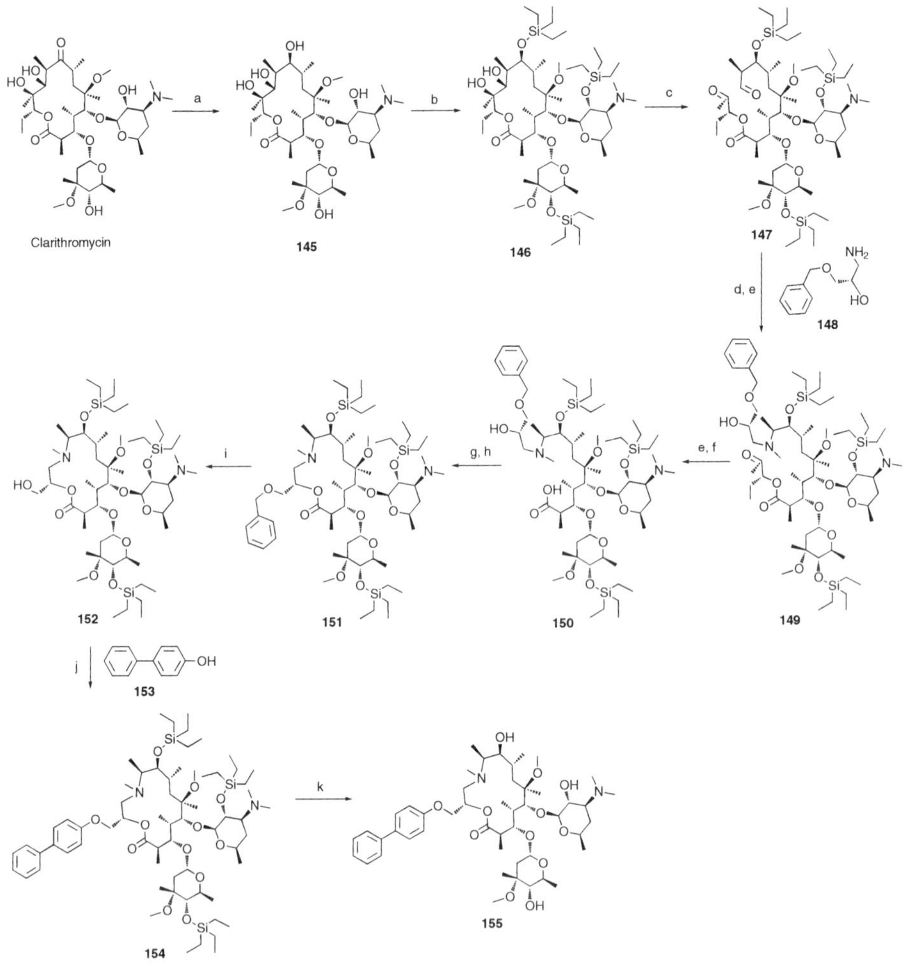

Scheme 2.27: The synthesis of 11a-azalides. Reagents and conditions: a) NaBH$_4$, ethanol (79%); b) TESCl, imidazole, DMF (96%); c) Pb(OAc)$_4$, DCM; d) **148**, NaBH(OAc)$_3$; e) HCHO (aq), NaBH(OAc)$_3$; f) LiOH, THF/ethanol/water; g) 2,4,6-Cl$_3$C$_6$H$_2$COCl, triethylamine, THF; h) DMAP, CH$_3$CN, reflux; i) H$_2$, Pd(OH)$_2$, THF (53%); j) **153**, DEAD, PPh$_3$, THF; k) HF–pyridine, THF (36% over two steps).

2.6 Conclusion

Overcoming resistance of bacteria towards existing macrolide antibiotics, as well as antibiotics in general, is an enormous challenge for researchers. However, the remarkable advancement in the development of semisynthetic methods that can be applied on chemically very complicated macrolide natural products assures the synthesis of novel rationally designed macrolide classes. Hopefully with further development of modern structural, computational and biological tools, novel macrolide-like molecules will be tailored against currently nonreachable targets like most Gram-negative and constitutively resistant bacterial strains. Unfortunately, the adverse effects of telithromycin, especially hepatotoxicity, have caused skepticism by drug regulatory agencies about the safety of the newer ketolides [51]. As increased clinical trial requirements and medium returns on investments decreased the interest of big pharmaceutical companies towards development of new macrolides, the main impact on the field could come from small pharmaceutical and biotech companies as well as from academia.

References

[1] Bunch RF, McGuire JM. Erythromycin, its salts, and method of preparation. US Patent 2,653,899 (1953).
[2] Doak BC, Over B, Giordanetto F, Kihlberg J. Chem Biol 2014, 21, 1115–1142.
[3] Kurath P, Jones PH, Egan RS, Perun TJ. Experientia 1971, 27, 362.
[4] Gasc J-C, D'Ambrieres SG, Lutz A, Chantot J-F. J Antibiot 1991, 44, 313–330.
[5] Morimoto S, Takahashi Y, Watanabe Y, Omura S. J Antibiot 1984, 37, 187–189.
[6] Watanabe Y, Morimoto S, Adachi T, Kashimura M, Asaka T. J Antibiot 1993, 46, 647–660.
[7] Watanabe Y, Adachi T, Asaka T, Kashimura M, Matsunaga T, Morimoto S. J Antibiot 1993, 46, 1163–1167.
[8] Morimoto S, Adachi T, Matsunaga T, Asaka T, Watanabe Y, Sota K, et al. Erythromycin A derivatives. US Patent 4,990,602, February 5, 1991.
[9] Clark RF, Ma Z, Wang A, Griesgraber G, Tufano M, Yong H, et al. Biorg Med Chem Lett 2000, 10, 815–819.
[10] Brunet E, Munoz DM, Parra F, Mantecon S, Juanes O, Rodriguez-Ubis JC, et al. Tetrahedron Lett 2007, 48, 1321–1324.
[11] Brunet E, Parra F, Mantecon S, Juanes O, Rodriguez-Ubis JC, Cruzado C, et al. Tetrahedron Lett 2008, 49, 2911–2915.
[12] Đokić S, Kobrehel G, Lazarevski G, Lopotar N, Tamburašev Z, Kamenar B, et al. J Chem Soc Perkin Trans I 1986, 1881–1890.
[13] Wilkening RR, Ratcliffe RW, Doss GA, Mosley RT, Ball RG. Tetrahedron 1997, 50, 16923–16944.
[14] Yang BV, Goldsmith M, Rizzi JP. Tetrahedron Lett 1994, 35, 3025–3028.
[15] Bright GM, Nagel AA, Bordner J, Desai KA, Dibrino JN, Nowakowska J, et al. J Antibiot 1988, 41, 1029–1047.
[16] Girard AE, Girard D, English AR, Gootz TD, Cimochowski CR, Faiella JA, et al. Antimicrob Agents Chemother 1987, 31, 1948–1954.
[17] Fiese EF, Steffen SH. J Antimicrob Chemother 1990, 25 (Suppl. A), 39–47.
[18] Đokić S, Kobrehel G, Lopotar N, Kamenar B, Nagl A, Mrvoš D. J Chem Res (S) 1988, 152–153.
[19] Bayod-Jasanada M, Carbajo RJ, Lopez-Ortiz F. J Org Chem 1997, 62, 7479–7481.
[20] Wilkening RR, Ratcliffe RW, Doss GA, Bartizal KF, Graham AC, Herbert CM. Bioorg Med Chem Lett 1993, 3, 1287–1292.
[21] Kobrehel G, Lazarevski G, Đokić S, Kučišec-Tepeš N, Cvrlje M. J Antibiot 1992, 45, 527–534.
[22] Wadel ST, Santorelli GM, Blizzard TA, Graham A, Occi J. Bioorg Med Chem Lett 1998, 8, 549–554.
[23] Wadel ST, Santorelli GM, Blizzard TA, Graham A, Occi J. Bioorg Med Chem Lett 1998, 8, 1321–1326.
[24] Denis A, Agouridas C. Bioorg Med Chem Lett 1998, 8, 2427–2432.
[25] Allen NE. Antimicrob Agents Chemother 1977, 11, 669–674.
[26] LeMahieu RA, Carson M, Kierstead RW, Fern LM, Grunberg E. J Med Chem 1974, 17, 953–956.

[27] Agouridas C, Denis A, Auger J-M, Benedetti Y, Bonnefoy A, Bretin F, et al. J Med Chem 1998, 41, 4080–4100.

[28] Baker WR, Clark JD, Stephens RL, Kim KH. J Org Chem 1988, 53, 2340–2345.

[29] Fernandes PB, Baker WR, Freiberg LA, Hardy DJ, McDonald EJ. Antimicrob Agents Chemother 1989, 33, 78–81.

[30] Denis A, Agouridas C, Auger J-M, Benedetti Y, Bonnefoy A, Bretin F, et al. Biorg Med Chem Lett 1999, 9, 3075–3080.

[31] Wei X, You Q. Org Proc Res Dev 2006, 10, 446–449.

[32] Cao Z, Liu B, Liu W, Yao G, Liu W, Wang Q. J Chem Res 2013, 37, 107–109.

[33] Or YS, Clark RF, Wang S, Chu DTW, Nilius AM, Flamm RK, et al. J Med Chem 2000, 43, 1045–1049.

[34] Bhatia AW. Strategies leading to the synthesis of a novel ketolide antibiotic. In: Harmata M, ed. Strategies and Tactics in Organic Synthesis 2004, 5, 133–152.

[35] Plata DJ, Leanna MR, Rasmussen M, McLaughlin MA, Condon SL, Kerdesky FAJ, et al. Tetrahedron 2004, 60, 10171–10180.

[36] Zhanel GG, Hartel E, Adam H, Zelenitsky S, Zhanel MA, Golden A, et al. Drugs 2016, 76, 1737–1757.

[37] Evrard-Todeschi N, Gharbi-Benarous J, Gaillet C, Verdier L, Bertho G, Lang C, et al. Biorg Med Chem 2000, 8, 1579–1597.

[38] Pereira D, Patel MK, Deo C. Process for the preparation of macrolide antibacterial agents. WO 2009/055557 A1, April 30, 2009.

[39] Fernandes P, Martens E, Pereira D. J Antibiot 2017, 70, 527–533.

[40] Kashimura M, Asaka T, Misawa Y, Matsumoto K, Morimoto S. J Antibiot 2001, 54, 664–678.

[41] Xu G, Tang D, Gai Y, Wang G, Kim H, Chen Z, et al. Org Process Res Dev 2010, 14, 504–510.

[42] Wang G, Niu D, Qiu Y-L, Phan LT, Chen Z, Polemeropoulos A, et al. Org Lett 2004, 6, 4455–4458.

[43] Čipčić-Paljetak H, Tomašković L, Matijašić M, Bukvić M, Fajdetić A, Verbanac D, et al. Curr Top Med Chem 2017, 17, 1–22.

[44] Fajdetić A, Čipčić Paljetak H, Lazarevski G, Hutinec A, Alihodžić S, Đerek M, et al. Bioorg Med Chem 2010, 18, 6559–6568.

[45] Kapić S, Čipčić Paljetak H, Alihodžić S, Antolović R, Eraković Haber V, Jarvest RL, et al. Bioorg Med Chem 2010, 18, 6569–6577.

[46] Palej Jakopović I, Kragol G, Forrest AK, Frydrych CSV, Štimac V, Kapić S, et al. Bioorg Med Chem 2010, 18, 6578–6588.

[47] Seiple IB, Zhang Z, Jakubec P, Langlois-Mercier A, Wright PM, Hog DT, et al. Nature 2016, 533, 338–345.

[48] Sugimoto T, Tanikawa T. ACS Med Chem Lett 2011, 2, 234–237.

[49] Sugimoto T, Tanikawa T, Suzuki K, Yamasaki Y. Bioorg Med Chem 2012, 20, 5787–5801.

[50] Deshpande PK, Sindkhedkar MD, Desai VN, Gupte SV, Yeole RD, Patel MV, et al. Azalides and azaketolides having antimicrobial activity. WO 2004/108744 A2, December 16, 2004.

[51] Metersky ML, Huang Y. Ann Res Hosp 2017, 1–4.

Predrag Novak

3 Interactions of macrolides with their biological targets

Molecular interactions play a crucial role in the processes of molecular recognition such as formation of ligand-protein, protein-protein or protein-nucleic acids complexes. Biologically relevant mechanisms are in most cases based on specific interactions, and therefore it is of the utmost importance in drug design to elucidate and understand these interactions.

Early steps in the process of drug discovery include identification of structural elements and groups that are responsible for bioactivity. However, our capability to design novel drug candidates only from high-resolution biomolecular structures is still limited and not straightforward. A thorough understanding of the molecular mechanisms and dynamics involved in the interaction of ligands with macromolecules is of crucial significance. This is one of the key prerequisites for the discovery of high affinity ligands for biologically important macromolecular targets.

There are numerous methods for studying ligand-receptor interactions, such as fluorescence spectroscopy, equilibrium dialysis, capillary electrophoresis, RNA footprinting, ultrafiltration, calorimetry, etc. Generally, many require time-consuming separation steps that might affect binding equilibria or derivatization steps, which may change ligand activity. The two most powerful and frequently used methods to study ligand-receptor complexes in modern drug discovery and development are NMR spectroscopy and X-ray diffraction methods. The two methods have advantages and disadvantages but should rather be considered as complementary. The main disadvantage of NMR spectroscopy is its inability to determine the structure of large biological systems (> 70 kDa) such as proteins, nucleic acids and their complexes, while crystallography can give precise structural data providing proper crystals are obtained. However, static structures obtained in the solid state may not be the same as the ones found in solution where a dynamic averaging of conformationally flexible structures is usually observed. However, by using numerous one- and two-dimensional NMR techniques, a plethora of information on ligand-receptor structure, interactions and properties can be obtained in solution including the mobility and dynamics of the studied systems.

Many NMR sequences have been developed in the last couple of decades, providing valuable data on the interactions between biological targets and small molecules that can be used to screen for potential small molecule binders [1–8].

Knowledge of the three-dimensional (3D) structures and conformational properties of ligand-receptor complexes might accelerate the discovery of more bioactive

Department of Chemistry, Faculty of Science, University of Zagreb, Horvatovac 102a, 10000 Zagreb, Croatia

https://doi.org/10.1515/9783110515756-003

compounds; however, optimization of physico-chemical properties such as ADMET (absorption, distribution, metabolism, excretion and toxicity), which dictate *in vivo* potency and efficacy, play an important role as well.

Macrolide antibiotics are effective and well-tolerated therapeutic agents for treating infectious diseases owing to their high efficacy and safety [9, 10]. They have been in widespread clinical use for over 60 years and are effective against Gram-positive and certain Gram-negative microorganisms. Since the discovery of the naturally produced macrolide antibiotic erythromycin in the early 1950s [11], many new natural or semisynthetic macrolide compounds have been discovered [12–28]. As already mentioned previously in this book, macrolides exert their biological activity by binding to the 50S bacterial ribosome subunit at an early stage of the translation process. A bacterial ribosome represents a macromolecular machine where genetic code is translated to proteins. Macrolides bind to the ribosomal 23S rRNA in domain V at or near the peptidyl transferase region and block the exit tunnel through which the nascent peptides leave the ribosome [29–31]. However, despite a number of existing antibiotics, the emerging multi-drug resistant microbial pathogens present serious and challenging problems in medical treatment that demand novel and more effective antimicrobial agents to be discovered. The WHO has recently expressed great concern that even minor injuries could become lethal owing to the fact that there are no efficient weapons to fight resistant pathogens.

In the year 2015, only one out of 45 newly approved drugs by the FDA was antibacterial, which is particularly concerning. Moreover, in the last 10 years, only nine new antibiotics have been approved and launched, which is significantly less than decades before and certainly not enough to overcome resistance.

There are four recognized mechanisms of bacterial resistance to macrolide antibiotics, e.g. modification in the ribosome target, efflux, inactivation of the compound and mutations in the 23S rRNA and proteins L4 and L22, but the first two are the most common types [32]. Erm methyltransferases are responsible for developing macrolide, lincosamide and streptogramin (MLS) resistance mechanisms of the inducible (iMLS) or constitutive (cMLS) types by methylating or dimethylating adenine 2058 of 23S rRNA and thus sterically blocking the macrolide binding [33–35]. The second prominent type of resistance is an active efflux by which macrolides are pumped out of the bacterial cell mediated by *Mef* genes [36].

To fight resistance, medicinal chemists can choose between the two main strategies to successfully resolve this issue. One includes design of novel macrolide scaffolds and/or derivatives of the existing ones to deal with resistance mechanisms, and the other involves a search for novel inhibitors of bacterial proteins involved in developing resistance, such as methyl transferases, for example [37].

Recently, ribosome crystallography has made significant progress in elucidating the structure and function of the ribosome [38–43]. Crystal structures of ribosome-macrolide complexes have shed new light on the binding mechanisms and interactions of macrolides to ribosomes and hence provide a good basis for the rational design of new antibiotics. Crystal structures of ribosome-macrolide complexes have been discussed in the previous chapter. A crucial step in the discovery of novel compounds for preventing resistance is to understand the principles of how macrolides interact with their macro-molecular targets [44, 45]. However, one should bear in mind the fact that crystal struc-tures are just static representations of a dynamic process involving different kinds of interactions and the mobility of the formed complex. They give an artificial state "frozen" in time and space and could reveal just one out of many possible conformational states. The real picture of macrolide-ribosome interactions is therefore more complex.

An important step in the design of potential inhibitors is to explore their interac-tions with biological targets and to elucidate bound conformations in solution. Hence, by applying one-and two-dimensional NMR experiments it is possible to characterize the interactions of macrolides and ribosomes and to determine the reactive groups responsible for binding. The two most commonly used NMR techniques to study mac-rolide binding are transferred nuclear Overhauser effect spectroscopy (trNOESY) and saturation transfer difference (STD) NMR experiments.

When the small ligand molecule binds to its target, the NOEs undergo remarkable changes leading to the detection of transferred nuclear Overhauser enhancements (trNOEs) arising from different correlation times of free and bound molecules. Small molecules exhibit short correlation times and give rise to positive NOEs or no NOEs at all. Large biomolecules, however, display negative NOEs. If a small molecule binds, it behaves like a macromolecule and shows strong negative trNOEs (Fig. 3.1).

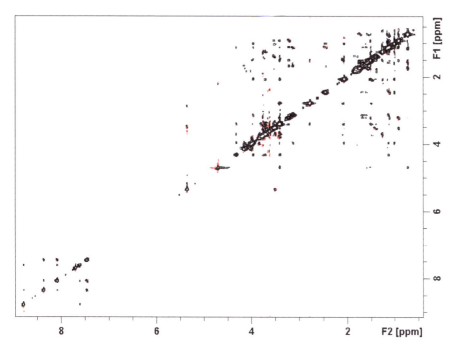

Fig. 3.1: trNOESY NMR spectrum of the antibiotic telithromycin bound to an *E. coli* ribosome.

STD NMR spectroscopy [46, 47] is based on the transfer of saturation from the biological receptor to the bound ligand *via* a spin-diffusion mechanism. Subsequently, the saturation is carried into solution by the exchange process and detected. The degree of saturation of the individual ligand resonances depends on the binding kinetics and the distance of the protons from the receptor. When the ligand dissociates, the saturation is transferred to the solution where the signals with narrow line widths are observed (Fig. 3.2). The main advantage of the STD technique is that there is no limitation on the size of the receptor so it is perfectly suited for ribosomes as well, and it requires only a small amount of the receptor molecule. It also benefits from the fact that only bound ligands display resonances in the spectrum and therefore could be used for screening purposes, too. STD spectroscopy can provide information on the groups of ligand molecule in intimate contact with its target, enabling a binding epitope to be determined as shown for azithromycin in Fig. 3.3.

Both techniques can be used to study macrolide interactions providing a fast exchange regime between the bound and free states. It has been shown that two interactions between macrolides and ribosomes exist in an allosteric two-stage binding process, e.g. a weak and a strong binding [48–52]. The fast-exchange weak binding

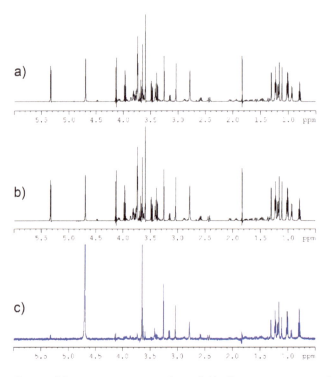

Fig. 3.2: a) On resonance spectrum (8 ppm), b) off resonance spectrum (50 ppm) and c) STD spectrum of an oleandomycin derivative recorded after addition of an *E. coli* ribosome in TRIS buffer at 25 °C.

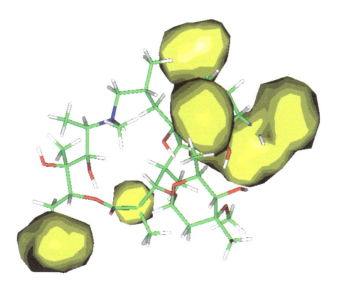

Fig. 3.3: Binding epitopes of azithromycin as determined by STD NMR spectroscopy.

seems to be the first step of recognition and selection of antibiotics by the ribosome and hence is suitable for NMR spectroscopic studies. The second step involves strong binding responsible for inhibition of bacterial protein synthesis. The weak binding was shown to be a necessary prerequisite for strong interactions [53] as the compounds that were not able to bind weakly did not exert antibiotic activity [54, 55].

Some early NMR studies of erythromycin-ribosome interactions have demonstrated that the ketone form is able to bind to *Escherichia coli* ribosomes [55, 56]. Titration experiments indicated a line broadening of signals belonging to protons involved in binding. A bound macrolide conformation was then assessed by using a NOESY experiment performed on macrolide-ribosome complexes.

The transferred NOESY NMR technique has also been used to study conformations of weakly bound antibiotics erythromycin and roxithromycin and their methylated derivatives to bacterial ribosomes isolated from *E. coli* strains [56]. In a continuation of their investigations, the authors studied some ketolide-ribosome interactions, including antibiotic telithromycin [57]. Significant differences between telithromycin and other studied ketolides (RU 72366, RU 004) were noticed, stressing the potency of telithromycin in inhibiting protein synthesis. Changes in the position of the alkyl-aryl side chain in the bound state were indicated, which was an important observation. This moiety has been shown to increase the affinity of the ketolide scaffold for the ribosome by several hundred-fold, corroborating that it is an important pharmacophore [58]. However, different bound conformations of this side-chain have been reported in ribosome crystal structures of different bacterial species [39–42]. In the structure of telithromycin bound to the *E. coli* ribosome, the alkyl-aryl chain was stacked on the A752-U2609 base pair [42], a conformation not observed in structures

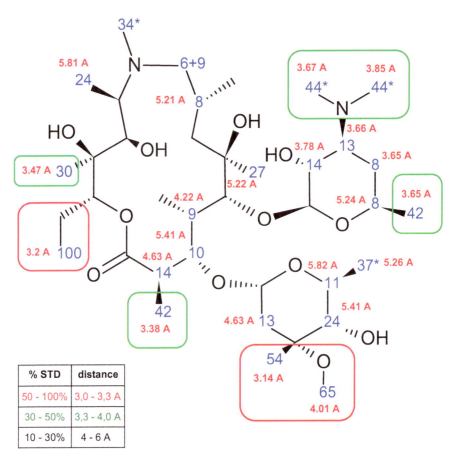

% STD	distance
50 - 100%	3,0 - 3,3 A
30 - 50%	3,3 - 4,0 A
10 - 30%	4 - 6 A

Fig. 3.4: Comparison of STD NMR signal enhancements and crystal structure distances for azythromycin [49].

obtained for *Haloarcula marismortui* and *Deinococcus radiodurans* ribosomes. It was folded back over the lactone ring when bound to the G2058A *H. marismortui* 50S subunit, or interacting with rRNA down the peptide exit tunnel when bound to the *D. radiodurans* 50S subunit.

The imidazolyl-pyridine chain in the solution-state weakly bound structures was reported to stretch to the right hand side of the macrocycle in the C-1 to C-3 region, stacking with carbonyl groups and laying between methyl groups at positions 15, 2 and 6 [57]. The end of this chain extended towards the desosamine sugar. Titration experiments enabled the identification of protons that were closest to the ribosome binding surface, including those at positions 12, 15, 2, 5 and some protons of the imidazolyl-pyridine chain.

Petropoulos and co-workers [59, 60] employed footprinting and kinetic methods to provide time-resolved details of the binding process of the macrolide antibiotics erythromycin, tylosin and azithromycin. Their findings confirmed previous results that macrolides bind to ribosomes in a two-step process: fast exchange low-affinity recognition binding located in the upper part of the exit tunnel and the slow high-affinity formation of a final complex also seen by crystallography. As previously mentioned, these facts inevitably show that a complete picture of the macrolide binding process is more complex than static crystal structures, which provide a snapshot of bound macrolide positions on the ribosome. These studies demonstrated that only one molecule of the drug was bound per *E. coli* ribosome at a time and that low- and high-affinity binding sites were mutually exclusive. Both interactions were found to be Mg^{2+} dependent. Azithromycin interacts initially with a hydrophobic crevice of *E. coli* ribosomes, formed by nucleosides G2057–A2059 at the upper part of the exit tunnel, being consistent with crystallographic data. The authors emphasized that species-specific structural differences may primarily account for the discrepancies between the antibiotic binding modes obtained for different organisms.

The free and bound structures of azithromycin and some homoerythromycin derivatives were studied by X-ray, NMR and molecular modeling calculations [61]. It was concluded that bound conformations were found to be very similar to those observed in the solution free state, being in agreement with the results obtained from the crystallography [39, 41]. The absence of a cladinose sugar unit was found to be the main cause of the inability of decladinosyl macrolides to bind to the ribosome. More detailed investigations employing a combination of STD NMR spectroscopy and *in vitro* protein inhibition on a series of macrolides have demonstrated that hydrophobic interactions involving a cladinose sugar, methyl groups of the lactone ring and the 13-alkyl moiety are a prerequisite for optimal positioning of the desosamine saccharide unit [49]. A synergy between the desosamine 3'-dimethylamino and 2'-hydroxyl interactions with the ribosome has been proposed as crucial for successful protein synthesis inhibition. These groups were also found by crystallography to be the reactive groups responsible for binding to the ribosome (Fig. 3.4).

Recently, trNOESY NMR experiments and molecular modeling were employed to determine the ribosome bound conformations of some 14- and 15-membered macrolides (erythromycin A, clarithromycin and azithromycin) and their decladinosyl derivatives [51]. This investigation demonstrated that three drugs adopted the "folded-out" conformation in the bound state, being in line with previously published data for azithromycin [50]. However, the bound conformations of these drugs were not completely superimposable. The clarithromycin bound conformation was found to be rigid with limited conformational flexibility, while azithromycin was more flexible when bound to the ribosome. However, modeling calculations pointed towards a range of conformations. The preferred conformation was the folded-out structure as indicated by NMR but a conformational flexibility was indicated owing to the increased size of the macrolide ring. The authors postulated that the chemical basis of the clinical differences between these antibiotics may be due to their different conformational flexibilities [51].

In order to assess and understand a complete biological profile of macrolides, it is also important to elucidate their interactions with other biological receptors such as cell membranes, bile acids or plasma proteins.

The overall bioactivity of a drug molecule depends also on its physico-chemical characteristics such as solubility, permeability and bioavailability. The latter is greatly influenced by binding to plasma proteins. Albumins belong to the class of plasma proteins that can reversibly interact with drugs and thus can serve as drug delivery systems. Bovine serum albumine (BSA) was used as a model system to study its interactions with azithromycin, oleandomycin and telithromycin [62]. Their binding epitopes as determined by employing STD NMR spectroscopy were found to be similar for azithromycin and oleandomycin, while some differences were noticed for telithromycin primarily in the alkyl-heteroaryl side chain and cladinose and desosamine sugar units.

One of the reasons for the success of macrolide antibiotics as drugs is their favorable physico-chemical profile and high accumulation in cells and tissues [63]. Furthermore, it has been reported that macrolides accumulate in lysosomal membranes and may induce phospholipidosis by inhibiting the activity of phospholipase A1 [64, 65]. Phospholipidosis is a lysosomal storage disorder characterized by the excess accumulation of phospholipids in tissues [66, 67]. Hence, to better understand these effects elucidation of macrolide-membrane interactions is desirable.

In that respect, interaction strength and localization of macrolide antibiotics with membrane-mimetics have been studied by Kosol and co-workers [45]. They used NMR translational diffusion and solvent paramagnetic relaxation enhancement (PRE) experiments to study binding of a series of macrolide compounds, including azithromycin, erythromycin A, azahomoerythromycin, clarithromycin, decladinosylazithromycin and azithromycin aglycone with dodecylphosphocholine (DPC) and sodium dodecylsulphate (SDS) micelles.

NMR experiments based on translational diffusion have already been proven to be useful to probe ligand-receptor interactions [68]. Information about the binding strength between macrolide antibiotic and bile-acid micelles were obtained from diffusion ordered NMR spectroscopy (DOSY) [69]. DOSY NMR spectrum is a pseudo-two-dimensional spectrum where chemical shifts represent one dimension while diffusion coefficients represent the other. When ligand binds to its target, its hydrodynamic radius increases dramatically, leading to a decrease in the diffusion rate and translational diffusion coefficients. Measured diffusion coefficients showed substantial differences between a free antibiotic and a mixture of antibiotic and bile acids, which confirmed the strong interaction and binding of the macrolides to bile micelles [69].

Paramagnetic relaxation enhancements (PREs) induced by an inert lanthanide complex can provide further information about the conformation, the mode of binding and the immersion depth of a drug molecule bound to its target as demonstrated for membrane-bound peptides [70, 71]. Upon the addition of the paramagnetic agent to the solvent, relaxation enhancements of the ligand nuclei depend on the insertion depth in the target. Providing the ligand is in the fast exchange between the free and bound state, solvent PREs are partly transferred to the free ligand and observed. Hence, by titrating the soluble and inert paramagnetic agent Gd(DTPA-BMA) into the solutions of macrolides bound to micelles the environment around the micelle was made paramagnetic, which led to solvent PREs depending on the distance to the surface of the micelle. All macrolide compounds showed a similar orientation to the micelles with the desosamine saccharide unit being closer to the surface than the cladinose unit and the right-hand-side macrolactone ring (protons 2, 3 and 4) extending deeper into the micelle, and the left-hand side being closer to the surface (protons 8, 9, and 10) [45]. This orientation enables electrostatic interaction between the positively charged amino groups of the macrolides with the negatively charged phosphate group of the phospholipid headgroup in micelles (Fig. 3.5). Earlier study by fluorescence spectroscopy also indicated binding of azithromycin close to the interfacial region of membrane-mimetics [72].

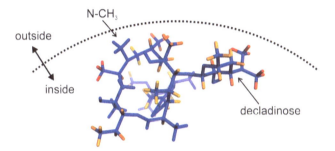

Fig. 3.5: Three-dimensional representation of the orientation of azithromycin in DPC micelles showing the protons for which solvent PREs could be obtained color-coded from red (high) to yellow (low). The localization and position relative to the membrane-mimetic is indicated and drawn to scale [45].

Proton NMR experiments corroborated that there was no direct interaction between macrolides and phospholipase A1, instead phospholipidosis was induced by the protection of lipids by macrolides, most likely by covering the membrane surface (Fig. 3.6).

As a continuation of these investigations the authors performed further studies on the same set of macrolides interacting with bile acids. It is known that about 50% of the total human drug clearance is made through the biliary pathway [73]. About 60%–70% of macrolide antibiotics were shown to be excreted by bile [74] and the rest by urine [75]. A tissue disposition study of azithromycin in rabbits has also shown that the highest tissue concentrations of this drug are found in bile [76]. Macrolide antibiotic uptake by bile has further been confirmed by experiments with bile duct cannulated rats, showing significantly reduced plasma concentrations of roxithromycin [77]. Also, they have been implicated in biliary system diseases such as cholestasis [78] and in modifying the biliary excretion of other drugs. For example, erythromycin

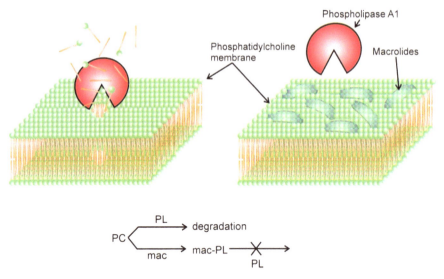

Fig. 3.6: Proposed mechanism of the protecting role of macrolides on lipid membranes. Instead of a direct binding to phospholipase A$_1$, the macrolides accumulate close to the surface of the membrane and thereby prevent access to the enzyme. PM, phosphatidylmembrane; PL, phospholipase A$_1$; mac, macrolide [45].

inhibits the excretion of ximelagatran and its metabolites [79] and reverses the bile salt tolerance in *Campylobacter jejuni* and *Campylobacter coli* strains [80].

The molecular details of interaction between macrolides and bile acids were investigated by NMR chemical-shift titration experiments, self-diffusion measurements, paramagnetic relaxation enhancements and small-angle X-ray scattering (SAXS), by using bile-acid micelles of cholic, deoxycholic and taurocholic acids as well as simulated intestinal fluids [69]. It has been demonstrated that studied macrolides bind to bile acids with different affinities. The strongest binding was observed between the most hydrophobic macrolides, azithromycin and clarithromycin, to cholate and deoxycholate micelles, respectively. The topology of these interactions as determined by solvent paramagnetic relaxation enhancements revealed that the macrolides were bound rather close to the surface of bile micelles, with no preferred orientation. Macrolides also interacted with mixed micelles of taurocholate-lecithin, but no binding to lecithin liposomes was detected. It was concluded that the interaction with bile did not impede macrolide antibiotics from targeting bacteria but enhanced the toxicity of bile and *vice versa* [69].

Competitive STD-NMR experiments were used to study the interactions of macrolide antibiotics and a glucosidase DesR [81], an enzyme involved in the development of a self-resistance mechanism of methymycin production [82, 83]. Data obtained suggested that the studied macrolides bind to the active site of the enzyme, C-9 to C-15 macrolide lactone moiety probably playing a critical role in the inhibitory effect. The authors claimed that these results could provide insights into the design of novel gly-cosidase inhibitors based on the macrolactone ring.

In conclusion, we are facing today bacterial strains that confer resistance to prac-tically all of the major antibiotic classes and it is clear that we need novel and more potent compounds to fight infections. Is this an uphill battle, one that we cannot win? Because no matter how hard we try, bacteria always fight back and find new ways to overcome antibiotic activity. We are all aware of the serious global threat to human health posed by growing bacterial resistance. There are only several tens of com-pounds in the current development pipeline to tackle serious resistant pathogens, which is certainly not enough.

Understanding how antibiotics exert their bioactivity and how they interact with biological macromolecules at the molecular level creates a good platform for medicinal chemists striving for novel solutions to fight infectious diseases. However, despite many new methods and approaches developed in the last several decades to understand mechanisms of antibiotic activity and bacterial resistance, some impor-tant pieces of the puzzle are missing and the macrolide antibiotic drug design is still scientifically challenging. A success in the discovery of new antimicrobial agents will crucially depend on further understanding of these interaction mechanisms, but also on changes in policy, marketplace and economic incentives as well as smoothing reg-ulatory pathways for anti-infectives to support and encourage antibiotic research and development. Researchers in the pharma industry and academia together with gov-ernment incentives should put much more effort into enhancing antibacterial drug discovery in the future.

References

[1] Stockman BJ, Dalvit C. NMR screening techniques in drug discovery and drug design. Prog Nucl Mag Res Sp 2002, 41, 187–231.
[2] Meyer B, Peters T. NMR spectroscopy techniques for screening and identifying ligand binding to protein receptors. Angew Chem Int Edit 2003, 42, 864–890.
[3] Pellecchia M, Sem DS, Wuthrich K. NMR in drug discovery. Nat Rev Drug Discov 2002, 1, 211–219.
[4] Tepeš P, Novak P. Ligand-receptor Interactions by NMR Spectroscopy. Kem Ind 2008, 57, 165–173.
[5] Coles M, Heller M, Kessler H. NMR-based screening technologies. Drug Discov Today 2003, 8, 803–810.
[6] Novak P, Jednačak T. NMR spectroscopy for studying interactions of bioactive molecules. In: Mandić Z., ed. Physico-chemical methods in drug discovery. Zagreb, IAPC Publishing, 2012, chapter 5, 189–231.
[7] Peng JW, Moore J, Abdul-Manan N. NMR experiments for lead generation in drug discovery. Prog Nucl Magn Reson Spectrosc 2004, 44, 225–256.
[8] Scwalbe H, Stilz HU, Kessler H. NMR spectroscopy of biomacromolecules in drug discovery and beyond. Chem Bio Chem 2005, 6, 1475–1478.
[9] Schönfeld W, Kirst HA. (eds.). Macrolide antibiotics. Basel, Birkhäuser Verlag, 2002.

[10] Pal S. A journey across the sequential development of macrolides and ketolides related to erythromycin. Tetrahedron 2006, 62, 3171–3200.

[11] McGuire JM, Bunch RL, Anderson RC, Boaz HE, Flynn EH, Powell HM, et al. Ilotycin, a new antibiotic. Antibiot Chemother 1952, 2, 281–283.

[12] Macrolide antibiotics: chemistry, biology, and practice, 2nd ed.; Omura, S. Ed. California, USA, Academic Press, 2002.

[13] Đokić S, Kobrehel G, Lazarevski G, Lopotar N, Tamburašev Z, Kamenar B, et al. Erythromycin series. Part 11. Ring expansion of erythromycin A oxime by the Beckmann rearrangement. J Chem Soc Perkin Trans 1 1986, 1881–1890.

[14] Đokić S, Kobrehel G, Lopotar N, Kamenar B, Nagl A, Mrvoš D. Erythromycin series. Part 13. Synthesis and structure elucidation of 10-dihydro-10-deoxo-11-methyl-11-azaerythromycin A. J Chem Research 1988, 152–153.

[15] Denis A, Agouridas C, Auger J-M, Benedetti Y, Bonnefoy A, Bretin F, et al. Synthesis and antibacterial activity of HMR 3647 a new ketolide highly potent against erythromycin-resistant and susceptible pathogens. Bioorg Med Chem Lett 1999, 9, 3075–3080.

[16] Or YS, Clark RF, Wang S, Chu DTW, Nilius AM, Flamm RK, et al. Design, synthesis, and antimicrobial activity of 6-O-substituted ketolides active against resistant respiratory tract pathogens. J Med Chem 2000, 43, 1045–1049.

[17] Bukvić Krajačić M, Dumić M, Novak P, Cindrić M, Koštrun S, Fajdetić A, et al. Discovery of novel ureas and thioureas of 3-decladinosyl-3-hydroxy 15-membered azalides active against efflux-mediated resistant *Streptococcus pneumoniae*. Bioorg Med Chem Lett 2011, 21, 853–856.

[18] Bukvić Krajačić M, Kujundžić N, Dumić M, Cindrić M, Brajša K, Metelko B, et al. Synthesis, characterization and *in vitro* antimicrobial activity of novel sulfonylureas of 15-membered azalides. J Antibiot 2005, 58, 380–389.

[19] Bukvić Krajačić M, Novak P, Cindrić M, Brajša K, Dumić M, Kujundžić N. Azithromycin–sulfonamide conjugates as inhibitors of resistant *Streptococcus pyogenes* strains. Eur J Med Chem 2007, 42, 138–145.

[20] Bukvić Krajačić M, Novak P, Dumić M, Cindrić M, Čipčić Paljetak H, Kujundžić N. Novel ureas and thioureas of 15-membered azalides with antibacterial activity against key respiratory pathogens. Eur J Med Chem 2009, 44, 3459–3470.

[21] Stodulkova E, Kuzma M, Hench IB, Cerny J, Kralova J, Novak P, et al. New polyene macrolide family produced by submerged culture of *Streptomyces durmitorensis*. J Antibiot 2011, 64, 717–722.

[22] Mandić Z, Naranđa A, Novak P, Brajša A, Đerek M. New derivatives of tylosin: chemical and electrochemical oxidation product of desmycosin. J Antibiot 2002, 9, 807–813.

[23] Sugimoto T, Tanikawa T, Suzuki K, Yamasaki Y. Synthesis and structure–activity relationship of a novel class of 15-membered macrolide antibiotics known as '11a-azalides'. Bioorg Med Chem 2012, 20, 5787–5801.

[24] Palej Jakopović I, Kragol G, Forrest AK, Frydrych CSV, Štimac V, Kapić S, et al. Synthesis and properties of macrolones characterized by two ether bonds in the linker. Bioorg Med Chem 2010, 18, 6578–6588.

[25] Vujasinović I, Marušić Ištuk Z, Kapić S, Bukvić-Krajačić M, Hutinec A, Đilović I, et al. Novel tandem reaction for the synthesis of N′-substituted 2-imino-1,3-oxazolidines from vicinal (sec- or tert-)amino alcohol of desosamine. Eur J Org Chem 2011, 13, 2507–2518.

[26] Zhao Z, Jin L, Xu Y, Zhu D, Liu Y, Liu C, et al. Synthesis and antibacterial activity of a series of novel 9-O-acetyl- 4′-substituted 16-membered macrolides derived from josamycin. Bioorg Med Chem Lett 2014, 24, 480–484.

[27] Shinde PB, Han AR, Cho J, Lee SR, Ban YH, Yoo YJ, et al. Combinatorial biosynthesis and antibacterial evaluation of glycosylated derivatives of 12-membered macrolide antibiotic YC-17. J Biotechnol 2013, 168, 142–148.

[28] Schieferdecker S, König S, Weigel C, Dahse HM, Oliver W, Nett M. Structure and biosynthetic assembly of gulmirecins, macrolide antibiotics from the predatory bacterium *Pyxidicoccus fallax*. Chem Eur J 2014, 20, 15933–15940.

[29] Tenson T, Mankin A. Antibiotics and the ribosome. Mol Microbiol 2006, 59, 1664–1667.

[30] Herman T. Drugs targeting the ribosome. Curr Opin Struct Biol 2005, 15, 355–366.

[31] Poehlsgaard J, Douthwaite S. The macrolide binding site on the bacterial ribosome. Curr Drug Targets Infect Disord 2002, 2, 67–78.

[32] Kwon JH. Macrolides, ketolides, lincosamides and streptogramins. In: Cohen J, Powderly WJ, Opal SM., ed. Infectious diseases. 4th ed. Elsevier, 2017, 1217–1229.

[33] Liu M, Douthwaite S. Resistance to the macrolide antibiotic tylosin is conferred by single methylations at 23S rRNA nucleotides G748 and A2058 acting in synergy. Proc Natl Acad Sci USA 2002, 99, 14658–14663.

[34] Hansen LH, Kirpekar F, Douthwaite S. Recognition of nucleotide G745 in 23S ribosomal RNA by the rrmA methyltransferase. J Mol Biol 2001, 310, 1001–1010.

[35] Weisblum B. Insights into erythromycin action from studies of its activity as inducer of resistance. Antimicrob Agents Chemother 1995, 39, 797–805.

[36] Chancey ST, Zhou X, Zähner D, Stephens DS. Induction of efflux-mediated macrolide resistance in *Streptococcus pneumoniae*. Antimicrob Agents Chemother 2011, 55, 3413–3422.

[37] Hajduk PJ, Dinges J, Schkeryantz JM, Janowick D, Kaminski M, Tufano M, et al. Novel inhibitors of Erm methyltransferases from NMR and parallel synthesis. J Med Chem 1999, 42, 3852–3859.

[38] Schlünzen F, Zarivach R, Harm J, Bashan A, Tocilj A, Albrecht R, et al. Structural basis for the interaction of antibiotics with the peptidyl transferase centre in eubacteria. Nature 2001, 413, 814–821.

[39] Hansen JL, Ippolito JA, Ban N, Nissen P, Moore PB, Steitz TA. The structures of four macrolide antibiotics bound to the large ribosomal subunit. Mol Cell 2002, 10, 117–128.

[40] Schlunzen F, Harms JM, Franceschi F, Hansen HAS, Bartels H, Zarivach R, et al. Structural basis for the antibiotic activity of ketolides and azalides. Structure 2003, 11, 329–338.

[41] Tu D, Blaha G, Moore PB, Steitz TA. Structures of MLSBK antibiotics bound to mutated large ribosomal subunits provide a structural explanation for resistance. Cell 2005, 121, 257–270.

[42] Dunkle JA, Xiong L, Mankin AS, Cate JHD. Structures of the *Escherichia coli* ribosome with antibiotics bound near the peptidyl transferase center explain spectra of drug action. Proc Natl Acad Sci 2010, 107, 17152–17157.

[43] Bulkley D, Innis CA, Blaha G, Steitz TA. Revisiting the structures of several antibiotics bound to the bacterial ribosome. Proc Natl Acad Sci 2010, 107, 17158–17163.

[44] Mankin AS. Macrolide myths. Curr Opin Microbiol 2008, 11, 414–421.

[45] Kosol S, Schrank E, Bukvić-Krajačić M, Wagner G, Meyer NH, Göbl C, et al. Probing the interactions of macrolide antibiotics with membrane-mimetics by NMR spectroscopy. J Med Chem 2012, 55, 5632–5636.

[46] Mayer M, Meyer B. Characterization of ligand binding by saturation transfer difference NMR spectroscopy. Angew Chem Int Ed 1999, 38, 1784–1788.

[47] Mayer M, Meyer B. Group epitope mapping by saturation transfer difference NMR to identify segments of a ligand in direct contact with a protein receptor. J Am Chem Soc 2001, 123, 6108–6117.

[48] Awan A, Brennan RJ, Regan AC, Barber J. Conformational analysis of the erythromycin analogues azithromycin and clarithromycin in aqueous solution and bound to bacterial ribosomes. J Chem Soc Chem Commun 1995, 1653–1654.

[49] Novak P, Barber J, Čikoš A, Arsić B, Plavec J, Lazarevski G, et al. Free and bound state structures of 6-*O*-methyl homoerythromycins and epitope mapping of their interactions with ribosomes. Bioorg Med Chem 2009, 17, 5857–5867.

[50] Novak P, Tatić I, Tepeš P, Koštrun S, Barber J. Systematic approach to understanding macrolide-ribosome interactions: NMR and modelling studies of oleandomycin and its derivatives. J Phys Chem A 2006, 110, 572–579.

[51] Arsic B, Awan A, Brennan RJ, Aguilar JA, Ledder R, McBain AJ, et al. Theoretical and experimental investigation on clarithromycin, erythromycin A and azithromycin and descladinosyl derivatives of clarithromycin and azithromycin with 3-*O* substitution as antibacterial agents. Med Chem Commun 2014, 5, 1347–1354.

[52] Bertho G, Gharby-Benarous J, Delaforge M, Lang C, Parent A, Girault JP. Conformational analysis of ketolide, conformations of RU 004 in solution and bound to bacterial ribosomes. J Med Chem 1998, 41, 3373–3386.

[53] Bertho G, Gharby-Benarous J, Ladam P, Delaforge M, Girault JP. Transferred nuclear Overhauser effect study of macrolide–ribosome interactions: correlation between antibiotic activities and bound conformations. Bioorg Med Chem 1998, 6, 209–221.

[54] Awan A, Brennan RJ, Regan AC, Barber J. The conformations of the macrolide antibiotics erythromycin A, azithromycin and clarithromycin in aqueous solution: a ^{1}H NMR study. J Chem Soc Perkin Trans 2, 2000, 1645–1652.

[55] Barber J, Gyi JI, Pye DA. Specific, weak binding of erythromycin A (ketone) and chloramphenicol to 50S subunits of *E. coli* ribosomes: a ^{1}H NMR study. J Chem Soc Chem Commun 1991, 1249–1252.

[56] Pye DA, Gyi JI, Barber J. Tautomeric recognition of erythromycin a by ribosomes: a ^{1}H nuclear magnetic resonance study. J Chem Soc Chem Commun 1990, 1143–1145.

[57] Evrard-Todeschi N, Gharbi-Benarous J, Gaillet C, Verdier L, Gertho G, Lang C, et al. Conformations in solution and bound to bacterial ribosomes of ketolides HMR 3647 (telithromycin) and RU 72366: a new class of highly potent antibacterials. Bioorg Med Chem 2000, 8, 1579–1597.

[58] Hansen LH, Mauvais P, Douthwaite S. The macrolide-ketolide antibiotic binding site is formed by structures in domains II and V of 23S ribosomal RNA. Mol Microbiol 1999, 31, 623–631.

[59] Petropoulos AD, Kouvela EC, Starosta AL, Wilson DN, Dinos GP, Kalpaxis DL. Time-resolved binding of azithromycin to *Escherichia coli* ribosomes. J Mol Biol 2009, 385, 1179–1192.

[60] Petropoulos AD, Kouvela EC, Dinos GP, Kalpaxis DL. Step-wise binding of tylosin and erythromycin to *Escherichia coli* ribosomes, characterized by kinetic and footprinting analysis. J Biol Chem 2008, 283, 4756–4765.

[61] Novak P, Banić Tomišić Z, Tepeš P, Lazarevski G, Plavec J, Turkalj G. Conformational analysis of oleandomycin and its 8-methylene-9-oxime derivative by NMR and molecular modelling. Org Biomol Chem 2005, 3, 39–47.

[62] Novak P, Tepeš P, Lazić V. Epitope mapping of macrolide antibiotics to bovine serum albumin by Saturation Transfer Difference NMR spectroscopy. Croat Chem Acta 2007, 80, 211–216.

[63] Stepanic V, Kostrun S, Malnar I, Hlevnjak M, Butkovic K, Caleta I, et al. Modeling cellular pharmacokinetics of 14- and 15-membered macrolides with physicochemical properties. J Med Chem 2011, 54, 719–733.

[64] Montene JP, Van Bambeke F, Piret J, Brasseur R, Tulkens PM, Mingeot-Leclercq MP. Interactions of macrolide antibiotics (erythromycin A, roxithromycin, erythromycylamine [Dirithromycin], and azithromycin) with phospholipids: computer aided conformational analysis and studies on a cellular and cell culture models. Toxicol Appl Pharmacol 1999, 156, 129–140.

[65] Reasor MJ, Kacew S. Drug-induced phospholipidosis: are there functional consequences? Exp Biol Med 2001, 226, 825–830.

[66] Anderson N, Borlak J. Drug-induced phospholipidosis. FEBS Lett 2006, 580, 5533–5540.

[67] Munic V, Banjanac M, Kostrun S, Nujic K, Bosnar M, Marjanovic N, et al. Intensity of macrolide anti-inflammatory activity in J774A.1 cells positively correlates with cellular accumulation and phospholipidosis. Pharmacol Res 2011, 64, 298–307.

[68] Lucas LH, Larive CK. Measuring ligand-protein binding using NMR diffusion experiments. Concepts Magn Reson Part: A 2004, 20, 24–41.

[69] Glanzer S, Pulido SA, Tutz S, Wagner GE, Kriechbaum M, Gubensäk N, et al. Structural and functional implications of the interaction between macrolide antibiotics and bile acids. Chem Eur J 2015, 21, 1–10.

[70] Respondek M, Madl T, Göbl C, Golser R, Zangger K. Mapping the orientation of helices in micelle-bound peptides by paramagnetic relaxation waves. J Am Chem Soc 2007, 129, 5228–5234.

[71] Zangger K, Respondek M, Göbl C, Hohlweg W, Rasmussen K, Grampp G, et al. Positioning of micelle-bound peptides by paramagnetic relaxation enhancements. J Phys Chem 2009, 113, 4400–4406.

[72] Tyteca D, Schanck A, Dufrene YF, Deleu M, Courtoy PJ, Tulkens PM, et al. The macrolide antibiotic azithromycin interacts with lipids and affects membrane organization and fluidity: studies on Langmuir–Blodgett monolayers, liposomes and J774 macrophages. J Membr Biol 2003, 192, 203–215.

[73] Kostrubsky VE, Strom SC, Hanson J, Urda E, Rose K, Burliegh J, et al. Evaluation of hepatotoxic potential of drugs by inhibition of bile-acid transport in cultured primary human hepatocytes and intact rats. Toxicol Sci 2003, 76, 220–228.

[74] Chelvan P, Hamilton-Miller J, Brumfitt W. Biliary excretion of erythromycin after parenteral administration. Br J Clin Pharmacol 1979, 8, 233–235.

[75] Zuckerman JM, Qamar F, Bono BR. Review of macrolides (azithromycin, clarithromycin), ketolids (telithromycin) and glycylcyclines (tigecycline). Med Clin North Am 2011, 95, 761– 791.

[76] Cárceles CM, Fernández-Varón E, Marín P, Escudero E. Tissue disposition of azithromycin after intravenous and intramuscular administration to rabbits. Vet J 2007, 174, 154–159.

[77] Lee JH, Park YJ, Oh J-H, Lee Y-J. Decrease in gastrointestinal absorption of roxithromycin in bile duct cannulated rats due to depletion of bile salts. Biopharm Drug Dispos 2013, 34, 360–364.

[78] Juricic D, Hrstic I, Radic D, Skegro M, Coric M, Vucelic B, et al. Vanishing bile duct syndrome associated with azithromycin in a 62-year-old man. Basic Clin Pharmacol Toxicol 2010, 106, 62–65.

[79] Matsson EM, Eriksson UG, Knutson L, Hoffmann K-J, Logren U, Fridblom P, et al. Biliary excretion of ximelagatran and its metabolites and the influence of erythromycin following intraintestinal administration to healthy volunteers. J Clin Pharmacol 2011, 51, 770–783.

[80] Mavri A, Možina SS. Resistance to bile salts and sodium deoxycholate in macrolide- and fluoro-quinolone-susceptible and resistant *Campylobacter jejuni* and *Campylobacter coli* strains. Microb Drug Resist 2013, 19, 168–174.

[81] Sadeghi-Khomami A, Lumsden MD, Jakeman DL. Glycosidase inhibition by macrolide antibiotics elucidated by STD-NMR spectroscopy. Chem Biol 2008, 15, 739–749.

[82] Zhao L, Sherman DH, Liu H-W. Biosynthesis of desosamine: molecular evidence suggesting β-glucosylation as a self-resistance mechanism in methymycin/neomethymycin producing strain, *Streptomyces venezuelae*. J Am Chem Soc 1998, 120, 9374–9375.

[83] Zhao L, Beyer NJ, Borisova SA, Liu H-W. β-Glucosylation as a part of self-resistance mechanism in methymycin/pikromycin producing strain *Streptomyces venezuelae*. Biochemistry 2003, 42, 14794–14804.

Federica Sodano[1], Maria Grazia Rimoli[2]

4 Hybrids of macrolides and nucleobases or nucleosides: synthetic strategies and biological results

4.1 Macrolide hybrid compounds: an excellent approach

As drug discovery is a time-consuming and costly process, a rational drug design involving modification of existing drugs is often the most affordable and productive approach. In line with this, the strategy of creating new hybrid scaffolds based on current medicines has emerged in recent years [1]. Hybrid molecules are defined as synthetic compounds containing two or more covalently linked pharmacophore moieties to cooperatively generate in a single molecule distinctive characteristics and/or biological properties. Considering the years-long success of the macrolide antibiotics in clinical use, in particular for the treatment of respiratory tract, skin, soft tissue and genital tract infections, as well as the macrolide scaffold's functionality prospects in antimalarial, anti-infective, antibacterial and anti-inflammatory areas, it is evident that such a class of molecules presents countless opportunities for creating a range of promising hybrid molecules.

According to some authors [2, 3], hybrid molecules can be classified as follows:
1. *Conjugates*, in which the molecular building blocks containing the pharmacophores are separated by a linker group not found in the individual parent molecules. The linkers in most conjugates are metabolically stable.
2. *Cleavage conjugates* have a metabolically cleavable linker designed to release two drugs that interact independently with each target.
3. *Fused hybrid molecules* have the linker decreased in such a way that the pharmacophores are essentially touching.
4. *Merged hybrids* have their molecular building blocks combined as common structural features; they overlap in the structures of the starting compounds, giving rise to reduced and simpler molecules [2].

So far, most of the macrolide hybrid compounds have been designed as *conjugates*, *fused hybrids* and *merged hybrids*. One of the recurring needs of this approach in the case of the macrolides was born from the continuous "fight" by pharmaceutical companies to combat the alarming development of resistance to these antibiotics.

1 Department of Drug Science and Technology, University of Turin, Turin, Italy
2 Department of Pharmacy, University of Naples "Federico II", Naples, Italy

https://doi.org/10.1515/9783110515756-004

4.2 Macrolide antibiotic resistance

The antibacterial mode of action of macrolide antibiotics is mediated by their binding to the 50S subunit of the bacterial ribosome. Although some differences in the building and orientation of macrolide and ketolide antibiotics within the peptide exit tunnel were observed, key features of their positioning are conserved across species. The consequence of macrolide binding to the ribosome is the inhibition of protein synthesis resulting in the arrest of bacterial growth. Macrolides are generally bacteriostatic against clinically important bacteria, but at high concentrations could be bactericidal against susceptible organisms. As macrolide binding sites overlap, the mode of action of macrolides corresponds with structurally dissimilar antibiotics lincosamides (L) and streptogramin (S_B), and they are jointly referred to as MLS_B antibiotics. Accordingly, the mechanisms of resistance involving target site modifications will simultaneously cause resistance to multiple antibiotics.

Bacterial resistance to macrolide antibiotics most often involves the alterations within the target site, mainly mutations in 23S rRNA and ribosomal proteins, as well as methylation of A2058, a key residue with which MLS_B antibiotics interact [4]. Monomethylation or demethylation in the N6 position of adenine A2058 (*E. coli* numbering) by methyl transferase enzymes encoded by the *erm* gene family, sterically negates possible hydrogen bond formation with the macrolides, thus decreasing their affinity for ribosomes [5, 6]. Expression of the *erm* genes and subsequent Erm methyltransferase activity can be inducible (triggered by the presence of an inducer – i.e. erythromycin or its derivative with an unmodified cladinose sugar) or constitutive (produced irrespective of the inducer exposure) thus resulting in an inducible or constitutive phenotype [7]. Macrolide efflux is another clinically relevant resistance mechanism, characterized by the presence of efflux pumps encoded by different genes in streptococci and staphylococci [8].

Nowadays bacterial resistance to antibiotics for this class has reached globally worrying levels so the quest for novel macrolide hybrid compounds active against resistant strains is critical [9, 10].

4.3 How and why to conjugate macrolides to nucleobases or nucleosides: introduction

Numerous and previous studies have revealed that the incorporation of heteroaromatic rings in macrolide antibiotics led to additional interactions with the ribosome thus enabling their activity against bacteria with *erm*-mediated resistance. An analogous strategy was applied in the synthesis of macrolide-nucleoside and macrolide-nucleobase conjugates in order to enhance binding affinity of macrolides to bacterial ribosomes by supplemental hydrogen bonding and/or electrostatic interactions.

4.3.1 Nucleobases or nucleosides conjugates with erythromycin and azithromycin

4.3.1.1 Introduction

In the literature, suitable reaction conditions have been described to link nucleobases and nucleosides to erythromycin A derivatives without causing undesired reactions in the macrolide substructures. Although the preliminary reported results are not exciting as far as bioactivity is concerned, in fact only a weak antibacterial activity was observed, these macrolide conjugates to nucleobases or nucleosides are important from a synthetic point of view, as the feasibility of different routes was established. Owing to the well-known instability of erythromycin derivatives in acidic media (formation of internal hemiacetals or acetals if the CO group at C9 has not been modified, quick hydrolysis of the L-cladinose moiety followed by a slower release of D-desosamine, translactonisation, etc.) and basic media (ring opening, retro-aldol reaction, etc.) [11], the essential point of Costa and coworkers' work [12] is the synthetic approach adopted to link the macrolides to nucleoside substructures by means of protocols that avoid deprotection steps or workups involving strong acidic or basic treatments.

4.3.1.2 Synthesis

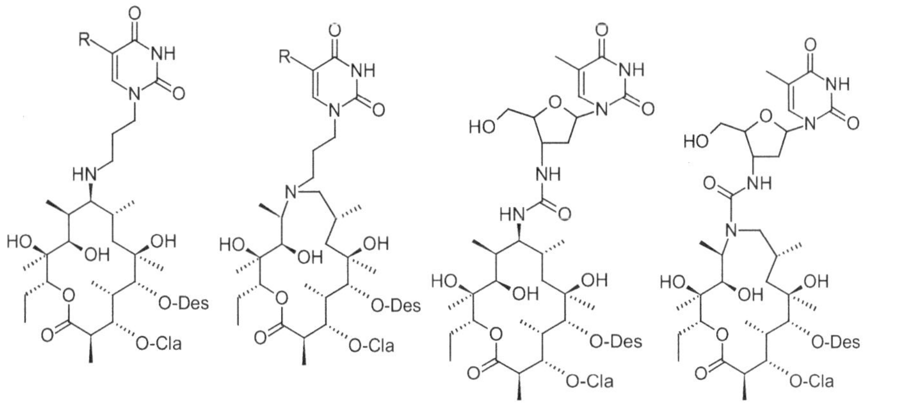

1a, R = Me
1b, R = H

2a, R = Me
2b, R = H

3

4

*Macrolide components of hybrids **1** and **3**.* Reduction of erythromycin A oxime (prepared from erythromycin A, EA, by using a very large excess of $HONH_3{}^+Cl^-$ in pyridine, at rt for 30 h) with $TiCl_3$ and $NaBH_3CN$, according to the procedure described by Leeds and Kirst [13], gave (9S)-erythromycylamine A (**5**). Treatment with acrolein (propenal) followed by *in situ* reduction with $NaBH_4$, as reported by Ryden et al. [14], afforded (9S)-9-N-(3-hydroxypropyl)erythromycylamine (**6**).

Macrolide components of hybrids 2 and 4. In two synthetic steps from EA oxime (Beckmann-like rearrangement followed by reduction), according to the procedure of Djokic et al. [15], it was prepared 9a-aza-9-deoxo-9a-homoerythromycin A (**7**), the synthetic precursor of the antibiotic azithromycin (*N*-Me-**7**). For the conversion of **7** into its *N*-3-hydroxypropyl derivative (**8**), the best results were obtained with $LiAlH_4$ in THF at 0 °C for 1 h.

Protected nucleobases as components of hybrids 1 and 2. Costa et al. used as nucleobases in their work thymine and uracil protected at N3 with Mocvinyl groups [16] because of their easy spectroscopic characterization and because this protecting group can be removed afterwards under very controlled conditions. The reaction of these nucleobases with methyl propynoate and DMAP, in CH_3CN at rt for 2 h, gave the bis-Mocvinyl derivatives **9a** and **9b**, respectively, in almost quantitative yields. To remove the group at N1, controlled amounts of morpholine or of morpholine plus DBU were employed under dilute conditions (0.01–0.05 M in CH_3CN) at rt; thus, **10a** [17] and **10b** were isolated by column chromatography, **9a** and **9b** being recovered in 20%–25%.

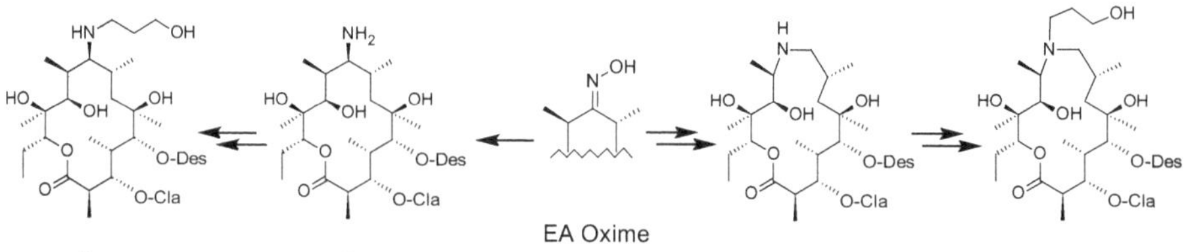

| 6 | 5 | EA Oxime | 7 | 8 |

Thymidine derivatives as components of hybrids 3 and 4. Reaction of amine **12** with *N,N'*-carbonyldiimidazole in CH_2Cl_2 at 0 °C gave **14**; this compound is stable and easy to purify.

Coupling of components. The Mitsunobu reaction between **6** and **10a**, in dioxane at rt for 1 h, afforded the desired Mocvinyl-protected **1a** [18]. It was supposed that secondary hydroxyl groups of L-cladinose and D-desosamine moieties would not significantly compete with the primary alcohol of **6**, as found. Deprotection to **1a** was accomplished in high isolated yield with pyrrolidine in CH_3CN, at rt for 24 h [19]. Similarly, **6** and **10b** gave **1b**, **8** and **10a** yielded **2a**, and **8** and **10b** gave **2b**.

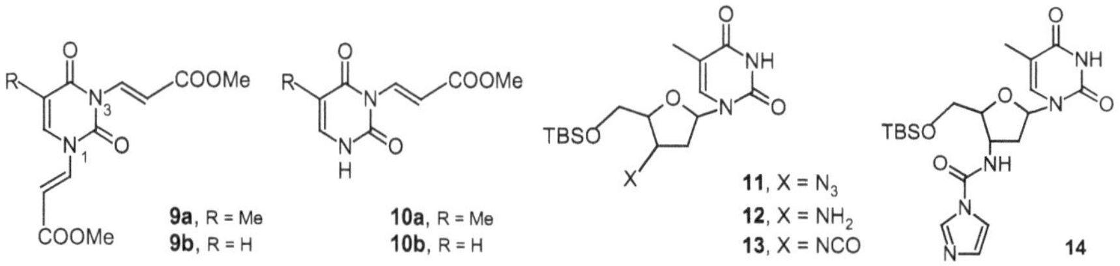

9a, R = Me
9b, R = H

10a, R = Me
10b, R = H

11, X = N₃
12, X = NH₂
13, X = NCO

14

Coupling of **6** with **14**, and that of **8** with **14**, which would give carbamate-linked hybrids, proceeded in refluxing CH_3CN complex mixtures [20]. On the other hand, coupling of amine **5** with **14** in CH_2Cl_2–DMF, at rt for 2 h, gave the urea, the TBS ether of which was cleaved with HF/pyridine in THF at 0 °C to afford hybrid **3**. Moreover, coupling of amine **7** with **14**, in CH_3CN–DMF at 50 °C for 2 h, afforded, also after deprotection with HF/pyridine, the desired urea **4** [21].

4.3.1.3 Biological results

The antibiotic activity of **1–4** has been screened against *Bacillus subtilis* ATCC6633, with a negative outcome (only between 4% and 13% of the activity of azithromycin). Compounds **1a, b** and **2a, b** were checked for their anti-mycobacterial activity (at TAACF, Birmingham, AL), showing 80%–83% inhibition against *Mycobacterium tuberculosis* H37Rv, and as an antitumoral (at NCI, Bethesda), but none was active enough to pass to the next step. Compound **4** showed no activity against the HIV-1 (NL4-3) 'wild-type' strain, nor against the HIV-1 AZT-resistant strain. The above-reported synthesis offers only starting points to design macrolide conjugates in the best way possible.

4.3.2 Clarithromycin-adenine and related conjugates

4.3.2.1 Introduction

As previously explained, erythromycin A (EA, **1a**) is unstable in the acidic media of the stomach and causes gastrointestinal side effects, so clarithromycin (**1b**) and roxithromycin (**1c**), a derivative of EA oxime (**1d**), were designed and synthesized; such semi-synthetic derivatives overcome this dangerous problem. The third generation macrolide antibiotics such as telithromycin [22–24] or ABT-773 [25] already use a similar strategy. Indeed, by incorporating heteroaromatic rings, they are capable of interacting further with the ribosome and enhancing the affinity for it. As reported previously, these rings appear to be crucial for activity against MSL_B resistant strains. Esteban and coworkers in their work [26] reported the synthesis of compounds **2–4**. In the series **2a–c**, the erythromycin backbone has been linked through an oxime ether to a nucleoside [27]; as a first approach, a thymidine was chosen. In clarithromycin derivatives, the link between the antibiotic and nucleoside (see **3a–c**) and between the antibiotic and adenine (see **4a, b**) has been established through a cyclic carbamate, or oxazolidone, at positions C11, C12 [28, 29].

4.3.2.2 Synthesis

For the preparation of **2** (Scheme 4.1), it is necessary to couple the thymidine derivative **5** prepared [12] from AZT with a suitably alkylated erythromycin oxime. Mesylates **6a–c** with different chain lengths were chosen as alkylating agents. Alkylation

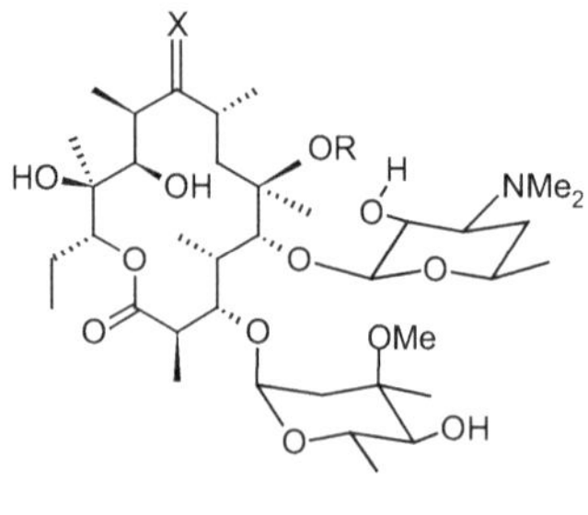

1a, R = H, X = O,   **EA**
1b, R = Me, X = O,   **Clarithromycin**
1c, R = H, X = NOMEM,   **Roxithromycin**
1d, R = H, X = NOH,   **EA oxime**

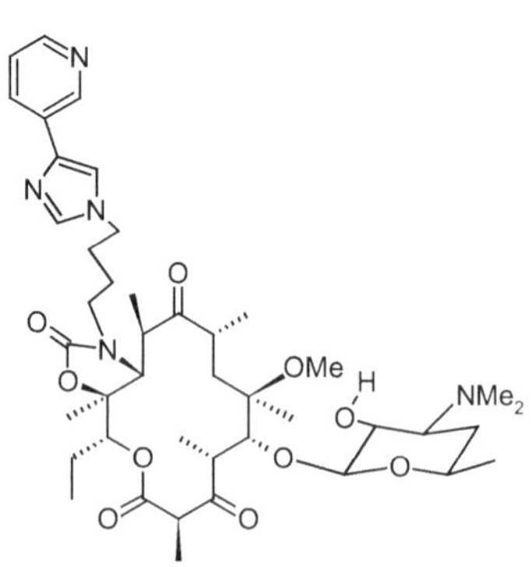

Telithromycin

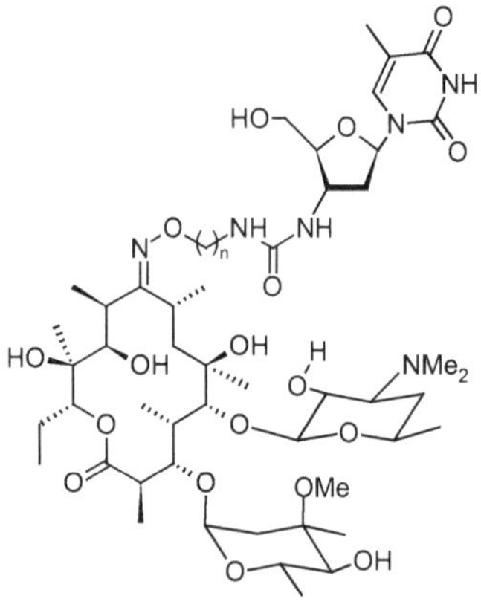

2

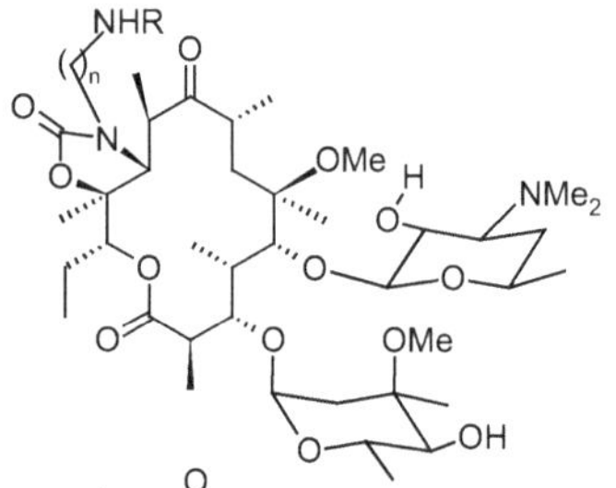

3, R =

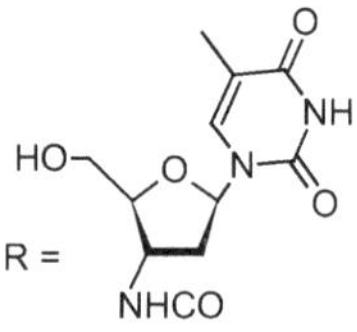

4, R = CH_2CH_2

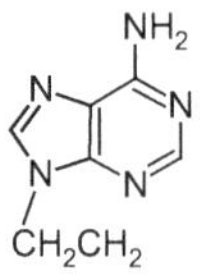

of the oxime **1d** was performed with K_2CO_3 as the base and acetone as the solvent; desired oximes (*E*)-**7a**–**c** were obtained in good yields. The corresponding (*Z*)-oximes, arising from the partial isomerization of EA (*E*)-oxime in the basic reaction medium [30], were also isolated. The *E* or *Z* configuration of **7a**–**c** was determined by careful comparison of their ^{1}H and ^{13}C NMR spectra with those of (*E*)- and (*Z*)-oximes of **1d** [30, 31], and **1b** [32]. The proton H11 appears at a lower field for the *Z*-oxime ethers due to the deshielding effect of the alkoxyimino group. The same effect causes H8 to be deshielded in the *E* compounds, in relation to the *Z* ones (Scheme 4.1). Furthermore, the ^{13}C NMR chemical shift for 10-Me is peculiar: it appears at ca. 15 ppm in the (*E*)-oxime ethers, and at ca. 11 ppm for the (*Z*)-oxime ethers. This is most probably due to a shielding effect caused by the steric interaction between the alkoxyimino group and 10-Me [30, 33].

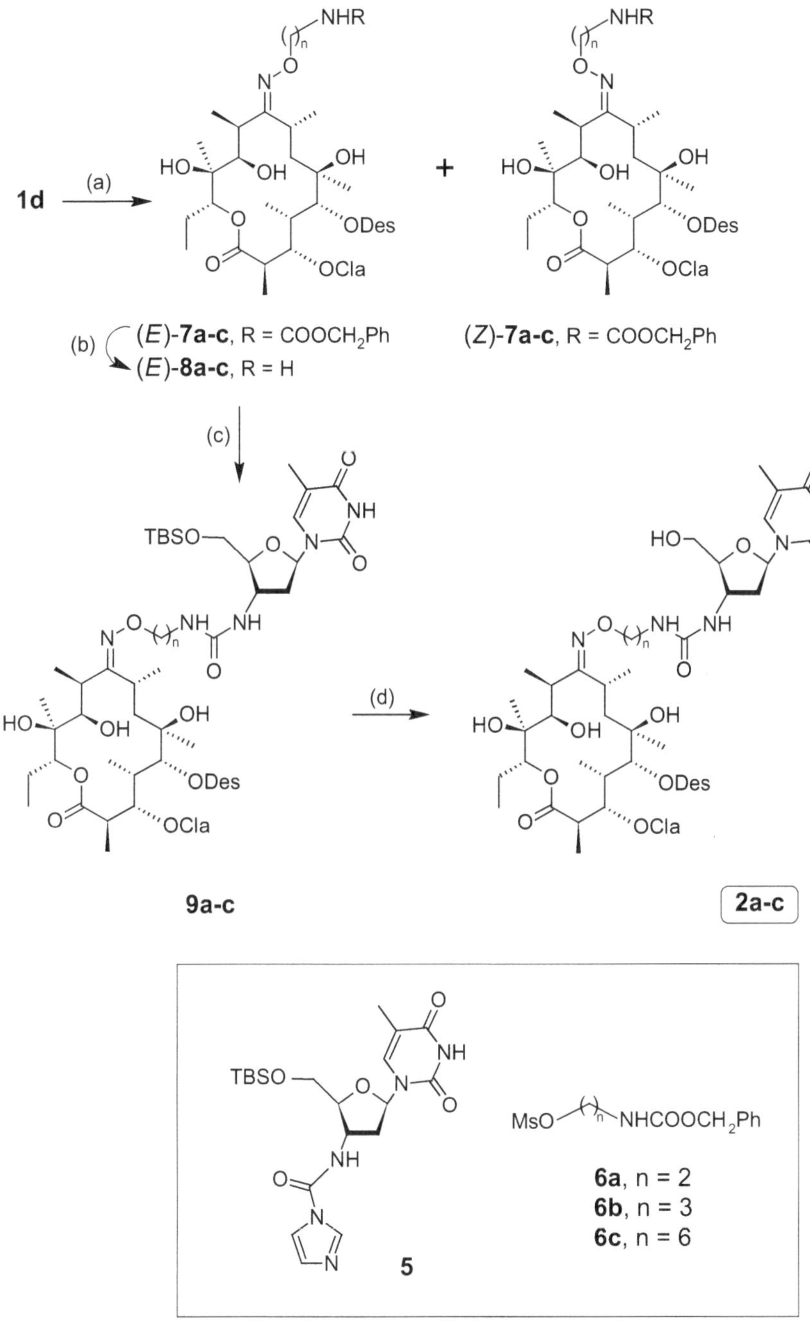
1d
(a)
(b)
(E)-7a-c, R = COOCH2Ph
(E)-8a-c, R = H
(Z)-7a-c, R = COOCH2Ph
(c)
(d)
9a-c
2a-c
5
6a, n = 2
6b, n = 3
6c, n = 6

Scheme 4.1

Removal of the benzyloxycarbonyl group of the major compounds, (E)-7a–c, by catalytic hydrogenation afforded the free amines 8a–c in quantitative yield. These amines were coupled with thymidine derivative 5 in CH_2Cl_2–DMF to yield protected chimeras 9a–c. Removal of the TBS group was accomplished by treatment with HF-pyridine in THF as the solvent to give the desired 2a–c.

Chimeras 3a–c were prepared by coupling amines 12a–c with thymidine 5, as shown in Scheme 4.2. The clarithromycin 2',4"-diacetate, 10, was converted to 12-O-imidazolylcarbonylmacrolide 11 by treatment with an excess of 1,1'-carbonyldi-imidazole and NaH in DMF–THF. This transformation proceeds through a 11,12-cyclic carbonate intermediate [28a]. The 11,12-carbamate group was introduced by treating crude 11 with the corresponding diamines followed by deprotection of the 2'-acetate. Only the natural 10R epimer was formed. For 12a–c, the stereochemistry at C10 was determined by NMR spectroscopy. Previous studies showed that the ^{13}C signal for C10 is diagnostic [28a]: it appears at ca. 37 ppm in the natural isomer, whereas it is shifted to ca. 52 ppm for the unnatural 10S epimer.

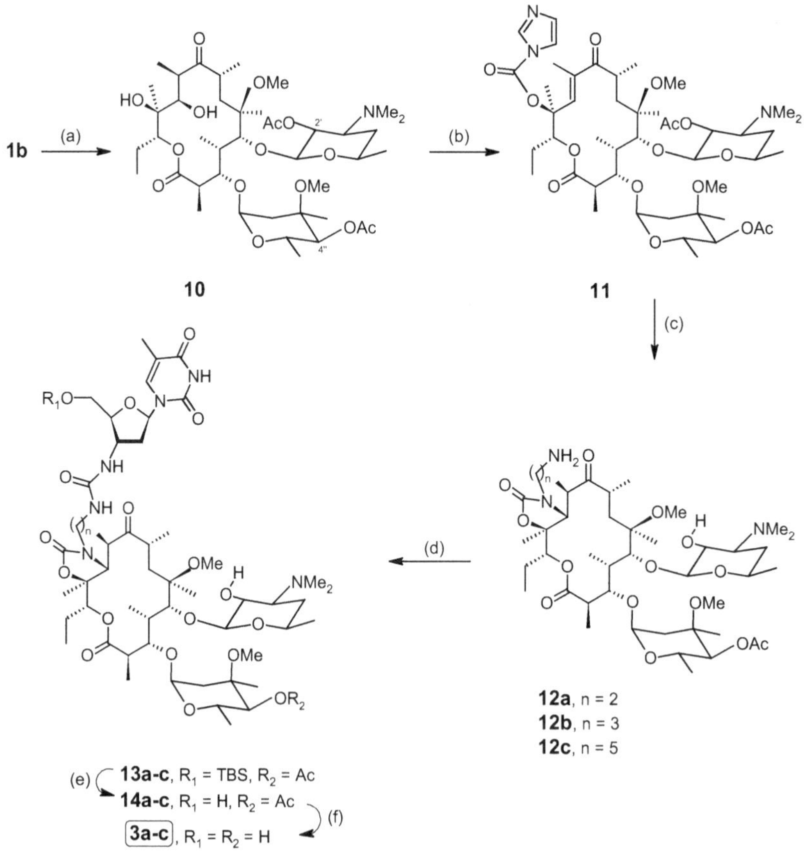

Scheme 4.2

Coupling of amines **12a–c** with **5** took place smoothly in CH$_2$Cl$_2$–DMF. Removal of the TBS group (to convert **13a–c** into **14a–c**, respectively), followed by treatment with aqueous ammonia to cleave the 4"-OAc groups, furnished the desired compounds **3a–c**.

Regarding macrolide–adenine chimeras **4a, b**, in which the link between the two components is through an amine moiety, they were prepared by reductive amination of amines **12a, b** with the hydrate of aldehyde **15** [34], followed by hydrolysis of the OAc group (Scheme 4.3).

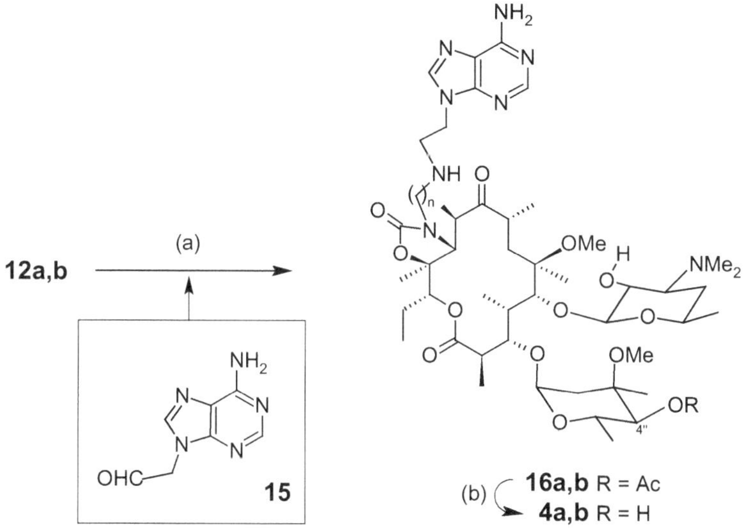

Scheme 4.3

4.3.2.3 Biological results

Compounds **2a–c**, **12a–c**, **13a–c**, **14a–c** and **3a–c** were tested against *Micrococcus luteus* ATCC 9341 by the standard agar dilution method [35]. All compounds showed weak antibacterial activity. The most active ones were amines **12a–c**, although with potencies below 40% of that of the reference antibacterial agent (EA, **1a**). Under identical conditions the value that we have determined for clarithromycin, **1b**, is 122% and that for azithromycin is 110%. Long-chain spacers (n = 6 for the case of **2c** and its precursors, n = 5 for the case of **3c** and its precursors) did not afford any advantage. All these side chains seem to be too bulky. Perhaps they also lack basic centres that could be required for the interaction with the nucleotides in the peptidyl-transferase tunnel [36].

Also, compounds **4a, b** and their precursors **16a, b** were tested against *M. luteus* ATCC 9341 [35]. Among all the samples tested so far, 4"-*O*-acetyl derivative **16a** showed the highest potency, although it is still lower than desired. A spacer with an additional CH$_2$ (from n = 2–3) is detrimental; a HB donor at 4" (4"-OH) is also detrimental regarding its 4''-OAc derivative. The antimycobacterial activities of compounds **2a–c** and **3a–c** were checked against *Mycobacterium tuberculosis* H37Rv [37]. Inhibition percentages around 73%–83% were noted.

4.4 Novel spiramycin-like conjugates: synthesis and antibacterial and anticancer evaluations

4.4.1 Introduction

Spiramycin is a natural antibiotic produced by *Streptomyces ambofaciens* in the form of a mixture of three compounds, the so-called spiramycins I–III [38, 39]. Spiramycin I, having a hydroxyl substituent at the C3 atom, is dominant (~80%) in the mixture produced by the bacteria [40]. Spiramycins are structurally similar to leucomycins in the presence of a common 16-membered lactone aglycone and mycaminosyl-mycarose moiety attached at C5. However, spiramycins, in contrast to leucomycins, additionally possess forosamine at C9 position. The spectrum of spiramycins' biological properties comprises mainly bacteriostatic activity against most Gram-(+) cocci and rods, mycoplasmas and *Toxoplasma gondii* [41, 42]. Comparison of spiramycins' activity against different Gram-(+) bacteria strains *in vitro* with that determined for 14-membered lactone macrolides such as erythromycins indicates that it is at least twice lower [43]. However, the advantage of spiramycins over erythromycins is their good gastrointestinal tolerance, higher affinity to tissues, and a fewer adverse effects [44, 45].

Spiramycin I, R = H
Spiramycin II, R = $COCH_3$
Spiramycin III, R = COC_2H_5

Klich K. and his coworkers [46] studied for the first time the activity of spiramycins in cancer cells, although the literature has already provided some examples of other-group macrolides showing properties of this type, e.g. maytansine (ansamacrolide) or marine 22-membered lactone macrolide (–)-dictyostatin [47–50]. The target site of spiramycins' action is well recognized from X-ray studies as the exit of the ribosomal tunnel near the peptidyl transferase loop and the loop of domain II of 23S rRNA, belonging to the large subunit 50S [7, 37, 51]. Spiramycin and related 16-membered lactone macrolides have a common mechanism of activity that involves binding to the 50S subunit and steric blocking of the peptide exit tunnel, which in turn contributes to inhibiting peptide synthesis at various stages. The stage of peptide synthesis inhibition is related to the length and chemical nature of the arm attached at the C5 position of the macrolide's aglycone [52]. As clarified above, bacterial resistance to 16-membered lactone macrolides is achieved when, e.g. nucleotides A2058 (*Escherichia coli*) or G2099 (*Haloarcula marismortui*) undergo N-methylation because of the steric clashing between the *N,N*-dimethyl group of the nucleotide and the mycaminose part [53, 54]. Among a whole group of mutations, the replacement of A2103 in *H. marismortui* ribosomes (A2062 in *E. coli*) seems to be the most important and confers resistance against spiramycin or tylosin [55]. Therefore, the modifications of the aldehyde group, which is usually reversibly bonded to the amine group of A2103 of the ribosome with formation of hemiaminal, for this type macrolides gave derivatives at least 100–1000-fold less active over unmodified parent antibiotics [56]. Up to now, many modifications, both within the spiramycin and leucomycin groups, have been proposed and tested to obtain comparably or more active alternative agents able to fight growing bacterial resistance. Most of the modifications were related to transformation of the aldehyde *via* reduction, nucleophilic addition [57–59]; hydroxyl groups within the aglycone and saccharides parts *via* etherification, acylation or sulphonylation [60, 61]; diene moiety *via* epoxidation, reduction, Diels–Alder reaction or metathesis with contraction of the lactone ring [62–65]; forosamine and/or mycaminose N-oxidation or N-substitution [66, 67] and incorporation of the nitrogen into the lactone macrocyclic ring [68–70]. An interesting platform for synthesis and drug discovery among the group of macrolide lactone antibiotics has been recently proposed by Seiple et al. [71]. A convenient synthetic strategy leading to the obtaining of new chemical entities characterized by attractive biological properties is the use of click chemistry reactions [72–74]. An interesting approach to modification of these structurally complexed macrolides, in view of their known mechanism of action, has been described by Omura and Sharpless et al. [75]. They proposed the functionalization of terminal hydroxyls of mycarose saccharide with alkyne followed by conversion into respective triazole moieties *via* Fokin–Huisgen cycloadditions. However, as a result of this transformation, the obtained derivatives were significantly less active than the parent leucomycin due to the presence of a too-long arm at C5 of the aglycone. After the novel cascade approach to modification of the aglycone of spiramycins, which opened a possibility of another pathway to construct the saccharide part at the aglycone [76], Klich et al., as a continuation of these studies, used a Huisgen cycloaddition to rebuild spiramycin's arm at C5 of the aglycone and to evaluate the influence of the modification on antibacterial and anticancer potency of novel triazole conjugates **6–16**.

4.4.2 Synthesis

To rebuild the spiramycin arm at C5, the earlier obtained compound **2** [76], *via* a sequence of cascade transformations from spiramycin, was further subjected to Huisgen dipolar cycloaddition with the use of different azides containing hydrophobic and hydrophilic substituents in the presence of CuOAc in THF/DMF mixture of solvents. Novel triazole conjugates of spiramycin's aglycone **6–16** were obtained with total yields 18%–60% after purification. The structures of all triazole conjugates **6–16** in solution were determined *via* 1D and 2D NMR and FT-IR spectroscopic analysis. The use of the HSQC method enabled the assignment of H23 and C23 of the triazole ring bonded to the aglycone structure. In turn, application of long-range heteronuclear couplings (HMBC) to structural analysis of conjugates **6–16** revealed simultaneous correlations of carbon and proton signals between the newly formed triazole ring and those of the tetrahydrofuran bicyclic moiety and R2 terminal substituent. Furthermore, the FT-IR spectrum of intermediate **2** shows two separated bands at 1722 and 1702 cm^{-1}, assigned to carbonyl stretching vibrations of aldehyde and double unsaturated lactone, respectively. In the spectra of novel triazole products **6–16**, one of these bands (v(C=O)aldehyde) vanishes, whereas the other one is shifted toward higher wavenumbers (to 1728 cm^{-1}). This result clearly indicated the lack of an aldehyde group within structures of **6–16**, and the fact that the lactone of the macrocyclic ring is no longer strongly π–π conjugated with the two double bonds in contrast to **2**. Additionally, for **6–10** and **16** conjugates, the presence of complex bands assigned to v(O–H) stretching vibrations confirmed the successful introduction of the saccharide part into the structure of new-type antibiotics. Application of ^{1}H–^{1}H NOESY method revealed a mutual arrangement of a newly introduced triazole ring relative to the modified aglycone part in solution. Generally, ^{1}H–^{1}H NOESY contacts recorded for **6–16**, assigned to the 16-membered aglycone part, can be divided into two groups, i.e. those assigned to the protons oriented above the aglycone and the others assigned to the protons oriented to the bottom of the aglycone ring. Taking into consideration the following group of contacts, H3–H6, H6–H11, H6–H19, H9–H11, H11–H13 and H13–H15, these protons are suggested to be at the same face of the aglycone (above the aglycone), whereas the second group of contacts (H2–H14, H2–H17, H5–H17, H10–H12, and H12–H14) is assigned to protons that are directed to the bottom of the aglycone. Mutual ^{1}H–^{1}H contacts between H5, H17 and H21 show their close vicinity and orientation of the –21CH$_2$– methylene protons to the bottom of the aglycone. As concluded from the presence of ^{1}H–^{1}H contacts found between H23 and H19 and H1′ in NOESY spectra, the arrangement of the triazole ring relative to bicyclic tetrahydrofuran moiety is similar for all obtained derivatives irrespective of the type of terminal substituent. Comparison of C5 arm structures between spiramycin and those of synthesized triazole conjugates **6–16** shows that they are of comparable length, whereas the mutual arrangement of C5 arms relative to the macrocyclic lactone ring is slightly different in each case.

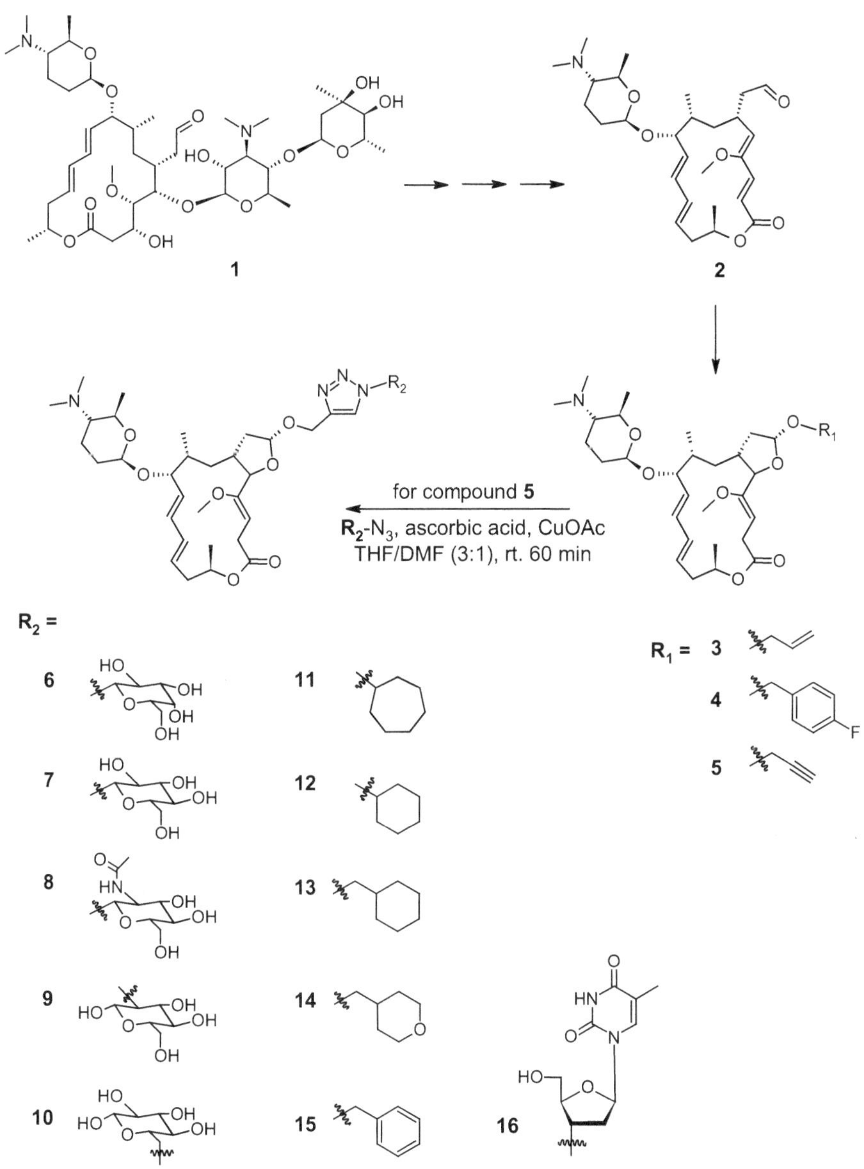

Scheme 4.4

4.4.3 Antibacterial studies

The analysis of antibacterial test results shows that **11–15** derivatives, containing bicyclic-triazole bridged aglycones, in general have no activity against all tested bacteria. First, the lack of an aldehyde group (no reversible covalent bonding) and limited possibility of stabilization of the terminal substituent at the C5 arm *via* H-bonds are the "weak points" of compounds **11–15** in view of the earlier published models [53, 77]. Additionally, although compounds **11–15** show the most favorable lipophilicity from among all other derivatives, they have the lowest solubility, which seems to be a very important factor considering the polar walls of the ribosomal tunnel and filling the tunnel with water molecules (transport to the target site of action). Compounds **3–5** also have poor solubility, and therefore they show no antibacterial activity. Similarly, low solubility of triazole conjugate containing AZT **16** can be an explanation for its lack of antibacterial activity. The importance of the role of reversible bonding of the aldehyde in the presence of antibacterial activity can be concluded from biological data of derivative **2**. The presence of antibacterial activity against Gram-positive microorganisms on the level 4–33 μM is mainly a result of forosamine interactions and the aldehyde bonding with the target site at ribosomes as well as quite well-balanced solubility and lipophilicity of compound **2**. However, it should be underlined here that the activity of **2** is about 8–28 times lower than that of the parent antibiotic **1**, when MICs are expressed in μM. This result is understandable because compound **2**, in contrast to **1**, does not have a saccharide part involved in hydrogen bonding with PTC loop nucleotides as G2099 and G2540 (*H. marismortui*). Interesting results were obtained for some of derivatives **6–10** containing terminal saccharides at the reconstructed C5 arm. Taking into account just their unfavorable lipophilicity and the best of all solubility parameters, no explanation of different antibacterial properties among them is possible, especially in view of **1**. This experimental result suggests slightly different binding modes among this group of triazole-saccharide conjugates. Further comparison of the physico-chemical parameters of compounds **6–10**, having a relatively hydrophilic C5 arm, shows that at good solubility of all these derivatives, just compound **8** has lipophilicity close to that of **1**, which contributes to its generally comparable activity with that of **1** (2–4 times less active than **1**). It should be underlined that conjugate **8** is also characterized by much higher activity than that of **2**, having an aldehyde group. These results clearly show that 16-membered macrolide derivatives, even those not containing an aldehyde group but having additionally functionalized saccharide, can be active against Gram-positive bacteria at a level comparable to that of **1**.

4.4.4 Anticancer studies

The leucomycin-type derivatives showed some anticancer activity [64]. However, the anticancer potency of spiramycin and its analogues were reported for the first time in Klich's work. Cytotoxic activity determined in six human cell lines (cervical (HeLa), nasopharyngeal (KB), breast (MCF-7), liver (HepG2), glioblastoma (U87), and normal human dermal fibroblast (HDF) of novel triazole conjugates), was compared to those of AZT, FUra, FdU, and ara-C standards. Analysis of these data revealed that **1** showed only limited cytotoxicity, irrespective of the type of the cancer cell line. Much lower activities (>100 µM) in comparison to that of the parent compound **1**, displayed derivatives **2**, **8**, and **9**. It should be mentioned here that the most active against Gram-positive bacteria, triazole conjugate **8**, has very weak anticancer activity. The other studied derivatives (compounds **10–16**) show significantly greater potency than that found for **1** in all cancer cell lines. Anticancer properties of the triazole conjugates contained within the C5 arm terminal saccharides (**8–10**) are varied. Derivative **10**, having 6′-substituted saccharide, revealed medium anticancer activity in HeLa, KB and U87 cell lines (IC_{50}~12 µM), in contrast to the other ones of this group, having saccharide attached *via* C1′ atom.

Incorporation of AZT into the C5-triazole arm resulted in increased activity of **16** in HeLa and KB cancer cell lines (IC_{50}~15 µM); however, lower than the activity of AZT itself. Interestingly, all triazole conjugates **11–15**, containing aliphatic or aromatic C5-terminal substituents, displayed interesting cytotoxic potency in all cancer cell lines. Among relatively hydrophobic triazole conjugates of the favorable log P values, the derivative bearing cycloheptyl substituent (compound **11**) is characterized by the highest potency in the range IC_{50} = 6.02–8.63 µM. IC_{50} values determined for **11** are only 2–3 times higher than those determined for cytarabine and are up to 5 times lower than for **1**. Considering the structures of **10–15** and their anticancer activity, it can be concluded that there is some relationship between the activity and the bulkiness of the C5-arm terminal substituent. The activity order beginning from the most active is the following: **11** with cycloheptyl >**13** with methylene cyclohexyl >**12** with cyclohexyl >**15** with benzyl group. Another favorable factor of novel hydrophobic triazole conjugates is their selectivity indexes (SI). Comparison of these indexes for **1** and all synthesized compounds revealed that those containing hydrophobic triazole arm (compounds **11–15**) have SI >1, whereas all others have SI ~1 or SI <1. A comparison of SI values determined for **11–15** with those of ara-C, FUra, and FdU standards demonstrates that the most active **11** has SI comparable or even higher than those of the standards, whereas also active compound **13** has even better SI than the standards used (except for HeLa cell line). It should be added that the most potent against bacteria, new derivative **8**, is 5-fold less toxic (less active in HDF cell line) than parent antibiotic spiramycin (compound **1**). Results of these studies clearly showed that for the anticancer activity of this group of macrolides, the presence of the aldehyde is not so important, whereas the occurrence of the hydrophobic and bulky structure of C5 arm is crucial. It has been demonstrated also that the use of combined cascade and click approaches to the modification of spiramycin allowed the obtaining of the new class of macrolide derivatives having interesting cytotoxic properties (comparable with FUra and even better than those of FdU).

References

[1] Meunier B. Hybrid molecules with a dual mode of action: dream or reality? Acc Chem Rev 2008, 41, 69–77.

[2] Morphy R, Rankovic Z. Designed multiple ligands. An emerging drug discovery paradigm. J Med Chem 2005, 48, 6523–6543.

[3] Muregi FW, Ishih A. Next-generation antimalarial drugs: hybrid molecules as a new strategy in drug design. Drug Dev Rev 2010, 71, 20–32.

[4] Paljetak HC, Tomaskovic L, Matijasic M, Bukvic M, Fajdetic A, Verbanac D, et al. Macrolide hybrid compounds: drug discovery opportunities in anti-infective and anti-inflammatory area. Cur Top Med Chem 2017, 17, 919–940.

[5] Douthwaite S, Jalava J, Jakobsen L. Ketolide resistance in *Streptococcus pyogenes* correlates with the degree of rRNA dimethylation by Erm. Mol Microbiol 2005, 58, 613–622.

[6] Katz L, Ashley GW. Translation and protein synthesis: macrolides. Chem Rev 2005, 105, 499–528.

[7] Gaynor M, Mankin AS. Macrolide antibiotics: binding site, mechanism of action, resistance. Curr Top Med Chem 2003, 3, 949–960.

[8] Leclercq R. Mechanism of resistance to macrolides and lincosamides: nature of the resistance elements and their clinical implication. Clin Infect Dis 2002, 34, 482–492.

[9] Holmes AH, Moore LSP, Sundsfjord A, Steinbakk M, Regmi S, Karkey A, et al. Understanding the mechanisms and drivers of antimicrobial resistance. Lancet 2016, 387, 176–187.

[10] Laxminarayan R, Matsoso P, Pant S, Brower C, Rottingen JA, Klugman K, et al. Access to effective antimicrobials: a worldwide challenge. Lancet 2016, 387, 168–175.

[11] For a classical review, see: (a) Omura S. Ed. Macrolide antibiotics: chemistry, biology, and practice. Orlando, Academic Press, 1984. For more recent, representative instances of the reactivity of erythromycins, see: (b) Sakan K, Babirad SA, Smith DA, Houk KN. Alkaline hydrolysis of some erythronolide a derivatives, Tetrahedron Lett 1990, 31, 3687–3690. (c) Bartra M, Urpí F, Vilarrasa J. An unexpected reaction in the lactamisation of 13-azido-13-deoxy-(9S)-9-dihydroerythronolide a seco-acid derivatives. Tetrahedron Lett 1992, 33, 3669–3672. (d) Waddell ST, Blizzard TA. Base catalyzed ring opening reactions of erythromycin A. Tetrahedron Lett 1992, 33, 7827–7830. (e) Pariza RJ, Freiberg LS. Erythromycin: new chemistry on an old compound. Pure Appl Chem 1994, 66, 2365–2368. (f) Faghih R, Edwards CM, Freiberg LA, Nellans HN. Entry into erythromycin lactams: synthesis of erythromycin A lactam enol ether as a potential gastrointestinal prokinetic agent. Synlett 1998, 751–753.

[12] Costa AM, Vilarrasa J. Hybrids of macrolides and nucleobases or nucleosides. Tetrahedron Lett 2000, 41, 3371–3375.

[13] Leeds JP, Kirst HA. A mild single-step reduction of oximes to amines. Synth Commun 1988, 18, 777–782.

[14] Ryden R, Timms GH, Prime DM, Wildsmith E. N-Substituted derivatives of erythromycylamine. J Med Chem 1973, 16, 1059–1060.

[15] Djokic S, Kobrehel G, Lazarevski G, Lopotar N, Tamburasev Z, Kamenar B, et al. Erythromycin series. Part 11. Ring expansion of erythromycin A oxime by the Beckmann rearrangement. J Chem Soc, Perkin Trans 1 1986, 1881–1890.

[16] (a) Faja M, Ariza X, Gálvez C, Vilarrasa J. Reaction of uridines and thymidines with methyl propyonate. A new N-3 protecting group. Tetrahedron Lett 1995, 36, 3261–3264. (b) Costa AM, Faja M, Farràs J, Vilarrasa J. Uracil- and thymine-substituted thymidine and uridine derivatives. Tetrahedron Lett 1998, 39, 1835–1838.

[17] Spectral data of 3-[(E)-2-(methoxycarbonyl)vinyl]thymine (**10a**): 1H NMR (DMSO-d_6, 200 MHz) δ1.80 (d, J = 1.0, 3H, C6-Me), 3.69 (s, 3H, COOMe), 6.95 (d, J = 14.7, 1H, CHCOOMe), 7.41 (dq,

J = 5.1, J = 1.2, 1H, H6), 8.18 (d, J = 14.6, 1H, CH=CHCOOMe), 11.29 (d, J = 5.1, 1H, NH); 13C NMR (DMSO-d_6, 50.3 MHz): δ12.6 (CH$_3$), 51.8 (CH$_3$), 107.4 (C), 110.7 (CH), 135.2 (CH), 137.6 (CH), 150.4 (CO), 163.2 (CO), 167.5 (CO); IR (KBr) ν1750, 1715, 1650; CIMS m/z 228 [M+NH$_4$]$^+$.

[18] To a suspension of **6** (71 mg, 0.090 mmol), Ph$_3$P (50 mg, 0.19 mmol), and **10a** (38 mg, 0.18 mmol) in dioxane (1–2 mL) was added dropwise DEAD (28 μL, 31 mg, 0.18 mmol) in 1 mL of dioxane. After stirring for 1 h, the solvent was evaporated and the products were separated by flash chromatography (from 95:5 CH$_2$Cl$_2$:MeOH to 95:5:2 CH$_2$Cl$_2$:MeOH:conc. ammonia) to give (9S)-9-N-{3-[3-(E)-2-(methoxycarbonyl)vinylthymin-1-yl]propyl}erythromycylamine A (76 mg, 87%).

[19] Pyrrolidine (13 μL, 11 mg, 0.16 mmol) was added to a solution of Mocvinyl-protected **1a** (39 mg, 0.04 mmol) in CH$_3$CN (0.4 mL). After stirring for 24 h, the solvent was removed and the residue was taken with CH$_2$Cl$_2$ and water. The aqueous layer was acidified to pH 5.0, the layers were separated, and the aqueous phase was rinsed with CH$_2$Cl$_2$. Afterwards, the aqueous solution was basified to pH 9.0 by addition of 2 M NaOH and was extracted three times with CH$_2$Cl$_2$. These organic extracts were collected and dried over anhydrous Na$_2$SO$_4$. Filtration and removal of the solvent *in vacuo* afforded chromatographically and spectroscopically pure (9S)-9-N-[3-(thymin-1-yl)propyl]erythromycylamine A, **1a** (30 mg, 83%): 1H NMR (CDCl$_3$, 300 MHz): δ0.88 (t, J = 7.5, 3H, Me-15), 1.01 (d, J = 6.5, 3H, Me-8), 1.10 (d, J = 6.5, 3H, Me-4), 1.11 (s, 3H, Me-12), 1.16 (d, J = 7.0, 3H, Me-10), 1.20 (d, J = 7.0, 3H, Me-2), 1.22 (d, J = 6.0, 3H, Me-50), 1.24 (s, 3H, Me-6 or Me-300), 1.26 (s, 3H, Me-300 or Me-6), 1.32 (d, J = 6.5, 3H, Me-500), 1.46–1.58 (m, 2H, H14a, H7b), 1.58 (dd, J = 15.3, J = 5.1, 1H, H200a), 1.67 (ddd, J = 12.6, J = 3.8, J = 2.1, 1H, H40b), 1.92 (d, J = 1.0, 3H, Me-thym.), 1.87–1.99 (m, 4H, H4, H14b, CH$_2$CH_2CH$_2$), 2.10–2.27 (m, 4H, H8, H9, H10, 400-OH), 2.29 (s, 6H, NMe$_2$), 2.38 (d, J = 15.0, 1H, H200b), 2.44–2.55 (m, 2H, H30, NHCHaHb), 2.71–2.79 (m, 1H, NHCHaHb), 2.89 (dq, J = 7.9, J = 7.0, 1H, H2), 3.03 (t, J = 8.7, 1H, H400), 3.25 (dd, J = 10.3, J = 7.1, 1H, H20), 3.32 (s, 3H, MeO-300), 3.50 (dqd, J = 10.6, J = 5.7, J = 1.8, 1H, H50), 3.58 (d, J = 7.5, 1H, H5), 3.67–3.83 (m, 2H, CH$_2$-thym.), 3.85 (br s, 1H, H11), 4.05 (dq, J = 9.1, J = 6.3, 1H, H500), 4.20 (d, J = 7.5, 1H, H3), 4.46 (d, J = 7.5, 1H, H10), 4.68 (dd, J = 10.2, J = 2.3, 1H, H13), 4.97 (d, J = 4.5, 1H, H100), 7.11 (q, J = 1.5, 1H, H6-thym.); 13C NMR (CDCl$_3$, 75.4 MHz): δ9.3 (CH$_3$), 11.1 (CH$_3$), 12.3 (CH$_3$), 15.2 (CH$_3$), 16.6 (CH$_3$), 16.8 (CH$_3$), 18.5 (CH$_3$), 21.4 (CH$_3$), 21.5 (CH$_2$), 21.5 (CH$_3$), 21.7 (CH$_3$), 26.6 (CH$_3$), 28.6 (CH$_2$), 29.3 (CH$_2$), 29.6 (CH), 31.2 (CH), 35.0 (CH$_2$), 36.5 (CH$_2$), 39.9 (CH), 40.2 (CH$_3$), 44.9 (CH), 45.9 (CH$_2$), 46.2 (CH$_2$), 49.4 (CH$_3$), 65.4 (CH), 65.5 (CH), 69.0 (CH), 70.4 (CH), 70.9 (CH), 71.1 (CH), 72.6 (C), 74.0 (C), 76.1 (C), 77.9 (CH), 78.1 (CH), 79.6 (CH), 83.8 (CH), 96.2 (CH), 103.2 (CH), 110.7 (C), 140.4 (CH), 150.8 (CO), 164.1 (CO), 177.7 (CO); HRFABMS, calcd for C45H81N4O14: 901.5749 [M+1]; found: 901.5726.

[20] Including the desired carbamate (FABMS 1176.7 [M+1]) but as a minor product. Reaction of **6** with crude isocyanate **13** gave similar mixtures.

[21] Spectral data of 9a-aza-9a-[30-deoxythymidin-30-yl)aminocarbonyl]-9-deoxo-9a-homoerythromycin A (**4**): 1H NMR (CD$_3$OD, 500 MHz) δ0.90 (t, J = 7.3, 3H, Me-15), 1.00–1.02 (m, 6H, Me-8, Me-4), 1.19–1.32 (m, 23H, H40a, H7a, Me-2, Me-6, Me-10, Me-12, Me-50, Me-300, Me-500), 1.42–1.54 (m, 2H, H14a, H7b), 1.58 (dd, J = 15.0, J = 5.0, 1H, H200a), 1.79–1.86 (m, 2H, H4, H40b), 1.85 (dqd, J = 14.3, J = 7.8, J = 2.3, 1H, H14b), 1.89 (d, J = 1.0, 3H, Me-thym.), 2.21 (br s, 1H, H8), 2.29–2.39 (m, 2H, H20-thym., H200-thym.), 2.42 (d, J = 16.0, 1H, H200b), 2.44 (br s, 6H, NMe$_2$), 2.80 (quint, J = 7.6, 1H, H2), 2.92 (t, J = 9.9, 1H, H30), 3.04 (d, J = 9.5, 1H, H400), 3.12 (br s, 1H, H9b or H10), 3.32 (s, 3H, MeO-300), 3.49 (d, J = 7.0, 1H, H5), 3.59 (br s, 1H, H11), 3.71 (dq, J = 9.4, J = 5.9, 1H, H50), 3.81 (dd, J = 12.5, J = 3.5, 1H, H50-thym.), 3.86 (dd, J = 12.0, J = 2.5, 1H, H500-thym.), 4.00 (br s, 1H, H40-thym.), 4.08 (d, J = 8.5, 1H, H3), 4.14 (dq, J = 9.5, J = 6.2, 1H, H500), 4.29 (q, J = 6.3, 1H, H30-thym.), 4.51 (d, J = 7.0, 1H, H10), 4.90 (d, J = 5.0, 1H, H100), 5.14 (d, J = 8.5, 1H, H13), 6.29 (t, J = 6.0, 1H, H10-thym.), 7.88 (br s, 1H, H6-thym.); 13C NMR (CD$_3$OD, 75.4 MHz): δ10.3 (CH$_3$), 11.8 (CH$_3$), 12.5 (CH$_3$), 13.4 (CH$_3$), 16.4 (CH$_3$), 18.7 (CH$_3$), 19.1 (CH$_3$), 20.8 (CH$_3$), 21.6 (CH$_3$), 21.8 (CH$_3$), 23.4 (CH$_2$), 29.0 (CH), 30.8 (CH$_2$), 31.7 (CH$_2$), 36.0 (CH$_2$),

39.1 (CH$_2$), 40.4 (CH$_3$), 41.9 (CH), 46.7 (CH), 50.0 (CH$_3$), 52.5 (CH), 53.6 (CH), 63.0 (CH), 63.0(CH$_2$), 65.6 (CH), 66.7 (CH), 68.9 (CH), 72.2 (CH), 74.3 (C), 76.1 (C), 76.4 (CH), 77.4 (C), 77.7 (CH), 79.2 (CH), 80.8 (CH), 85.9 (CH), 87.1 (CH), 87.2 (CH), 97.5 (CH), 104.3 (CH), 111.7 (C), 138.1 (CH), 153.2 (C), 160.8 (CO), 166.4 (CO), 178.1 (CO); HRFABMS, calcd for C48H85N5O17: 1003.5940 [M+1]; found: 1003.5900, calcd for C48H84N5O17: 1002.5862 [M]; found: 1002.5859.

[22] Agouridas C, Denis A, Auger J, Benedetti Y, Bonnefoy A, Bretin F, et al. Synthesis and antibacterial activity of ketolides (6-O-methyl-3-oxoerythromycin derivatives): a new class of antibacterials highly potent against macrolide-resistant and -susceptible respiratory pathogens. J Med Chem 1998, 41, 4080–4100.

[23] Denis A, Agouridas C, Auger J-M, Benedetti Y, Bonnefoy A, Bretin F, et al. Synthesis and antibacterial activity of HMR 3647 a new ketolide highly potent against erythromycin-resistant and susceptible pathogens. Bioorg Med Chem Lett 1999, 9, 3075–3080.

[24] Berisio R, Harms J, Schluenzen F, Zarivach R, Hansen HAS, Fucini P, et al. Structural insight into the antibiotic action of telithromycin against resistant mutants. J Bacteriol 2003, 185, 4276–4279.

[25] Or YS, Clark RF, Wang S, Chu DTW, Nilius AM, Flamm RK, et al. Design, synthesis, and antimicrobial activity of 6-O-substituted ketolides active against resistant respiratory tract pathogens. J Med Chem 2000, 43, 1045–1049.

[26] Esteban J, Costa AM, Cruzado MC, Faja M, Garcia P, Vilarrasa J. Clarithromycin-adenine and related conjugates. Tetrahedron Lett 2006, 47, 1919–1922.

[27] For reports of 9-oxime derivatives with good antibacterial activity, see: (a) Chantot JF, Bryskier A, Gasc JC. Antibacterial activity of roxithromycin: a laboratory evaluation. J Antibiot 1986, 39, 660–668; (b) Denis A, Pejac J-M, Bretin F, Bonnefoy A. Synthesis of 9-oxime-11,12-carbamate ketolides through a novel N-deamination reaction of 11,12-hydrazonocarbamate ketolide. Bioorg Med Chem 2003, 11, 2389–2394; (c) Kawashima Y, Yamada Y, Asaka T, Misawa Y, Kashimura M, Morimoto S, et al. Structure-activity relationship study of 6-O-methyleryth-romycin 9-O-substituted oxime derivatives. Chem Pharm Bull 1994, 42, 1088–1096; (d) Akemi N, Narita K, Ohmoto S, Takahashi Y, Yoshizumi S, Yoshida T, et al. Studies on macrolide antibiotics I. Synthesis and antibacterial activity of erythromycin A 9-O-substituted oxime ether derivatives against *Mycobacterium avium* complex. Chem Pharm Bull 2001, 49, 1120–1127.

[28] This ring is known to enhance antibacterial activity against resistant pathogens by increasing the conformational rigidity of the macrolactone. See: (a) Baker WR, Clark JD, Stephens RL, Kim KH. Modification of macrolide antibiotics. Synthesis of 11-deoxy-11-(carboxyamino)-6-O-methylery-thromycin A 11,12-(cyclic esters) *via* an intramolecular Michael reaction of O-carbamates with an α, β-unsaturated ketone. J Org Chem 1988, 53, 2340–2345; (b) Fernandes PB, Baker WR, Freiberg LA, Hardy DJ, MacDonald E. New macrolides active against *Streptococcus pyogenes* with inducible or constitutive type of macrolide-lincosamide-streptogramin B resistance. J Antimicrob Agents Chemother 1989, 33, 78–81.

[29] For ketolides with heteroaromatic appendages (including two examples with adenine) linked *via* a thioether, see: Hunziker D, Wyss P-C, Angehrn P, Mueller A, Marty H-P, Halm R, et al. Novel ketolide antibiotics with a fused five-membered lactone ring-synthesis, physicochemical and antimicrobial properties. Bioorg Med Chem 2004, 12, 3503–3519.

[30] Wilkening RR, Ratcliffe RW, Doss GA, Bartizal KF, Graham AC, Herbert CM. The synthesis of novel 8a-aza-8a-homoerythromycin derivatives *via* the Beckmann rearrangement of (9Z)-erythromycin A oxime. Bioorg Med Chem Lett 1993, 3, 1287–1292.

[31] (a) Esteban J, Costa AM, Urpí F, Vilarrasa J. From (E)- and (Z)-ketoximes to N-sulfenylimines, ketimines or ketones at will. Application to erythromycin derivatives. Tetrahedron Lett 2004, 45, 5563–5567; (b) Esteban J, Master Thesis, Universitat de Barcelona, 2002; (c) Wilkening RR. 9-Deoxo-9(Z)-hydroxy-iminoerythromycin A and O-derivatives thereof. EP 0503932 A1, September 16, 1992.

[32] Spectral data of clarithromycin (*E*)-oxime: 1H NMR (CDCl$_3$, 400 MHz): δ 0.84 (t, *J* = 7.4, 3H, Me15), 0.99 (d, *J* = 6.8, 3H, 8-Me), 1.08 (d, *J* = 8.0, 3H, 4-Me), 1.13 (s, 3H, 12-Me), 1.14 (d, *J* = 8.0, 3H, 10-Me), 1.20 (d, *J* = 6.8, 3H, 2-Me), 1.23 (d, *J* = 6.0, 3H, 5'-Me), 1.25 (s, 3H, 3''- Me), 1.31 (d, *J* = 6.4, 3H, 5''-Me), 1.48 (s, 3H, 6-Me), 1.48–1.61 (m, 4H, H7a+H7b+H14a+H2''a), 1.67 (m, 1H, H4'b), 1.89–1.98 (m, 2H, H4+H14b), 2.30 (s, 6H, NMe$_2$), 2.37 (d, *J* = 15.2, 1H, H2''b), 2.44 (m, 1H, H3'), 2.58 (q, *J* = 7.2, 1H, H10), 2.89 (dq, *J* = 9.2, *J* = 7.2, 1H, H2), 3.03 (m, 1H, H4''), 3.11 (s, 3H, 3''-OMe), 3.21 (dd, *J* = 10.0, 7.2, 1H, H2'), 3.33 (s, 3H, 6-OMe), 3.48 (m, 1H, H5'), 3.66 (d, *J* = 7.2, 1H, H5), 3.74–3.80 (m, 2H, H3+H8), 3.75 (s, 1H, H11), 4.03 (dq, *J* = 9.2, 6.4, 1H, H5''), 4.44 (d, *J* = 7.2, 1H, H1'), 4.94 (d, *J* = 4.4, 1H, H1''), 5.11 (dd, *J* = 11.2, 2.0, 1H, H13); 13C NMR (CDCl$_3$, 100.6 MHz): δ9.1 (4-Me), 10.6 (C15), 14.9 (10-Me), 16.0 (2-Me), 16.1 (12-Me), 18.6 (5''-Me), 18.7 (8-Me), 20.0 (6-Me), 21.2, 21.5, 21.5 (C14, 50 -Me, 3''-Me), 25.4 (C8), 28.8 (C4'), 32.9 (C10), 34.9 (C2'', 37.4 (C7), 39.1 (C4), 40.3 (NMe$_2$), 45.1 (C2), 49.5 (3''-OMe), 51.2 (6-OMe), 65.5 (C3'), 65.7 (C5''), 68.6 (C5'), 70.2 (C11), 71.1 (C2'), 72.7 (C3''), 74.1 (C12), 76.9 (C13), 78.0 (C4''), 78.5 (C6), 78.8 (C3), 80.5 (C5), 96.1 (C1''), 102.8 (C1'), 170.9 (C9), 175.7 (C1). The assignments indicated throughout have been established or corroborated by 2D NMR experiments (HSQC and COSY). (*Z*)-Oxime: 1H NMR (CDCl$_3$, 500 MHz): δ0.80 (t, *J* = 7.3, 3H, Me15), 1.0–1.4 (m, 28H, H4'a+2-Me+4-Me+6-Me+8-Me+1'-Me+12-Me+5'-Me+3''-Me+5''-Me), 1.40–1.50 (m, 2H, H7a+H14a), 1.51–1.57 (m, 2H, H7b+H2''a), 1.64 (m, 1H, H4'b), 1.86–1.95 (m, 2H, H4+H14b), 2.27 (s, 6H, NMe$_2$), 2.33 (d, *J* = 15.0, 1H, H2''b), 2.43 (ddd, *J* = 12.5, 10.5, 3.3, 1H, H30', 2.55 (m, 1H, H10), 2.73 (m, 1H, H8), 2.84 (dq, *J* = 9.3, 7.3, 1H, H2), 2.98 (d, *J* = 9.0, 1H, H4''), 3.06 (s, 3H, 3''-OMe), 3.18 (dd, *J* = 10.3, 7.3, 1H, H2'), 3.33 (s, 3H, 6-OMe), 3.44 (m, 1H, H5'), 3.58 (d, *J* = 7.5, 1H, H5), 3.73 (d, *J* = 9.5, 1H, H3), 3.93 (br s, 1H, H11), 3.98 (dq, *J* = 9.3, 6.3, 1H, H5''), 4.39 (d, *J* = 7.0, 1H, H1'), 4.89 (d, *J* = 4.5, 1H, H1''), 5.04 (dd, *J* = 11.0, 2.0, 1H, H13); 13C NMR (CDCl$_3$, 75.4 MHz): δ 9.1 (4-Me), 10.6 (C15), 11.6 (10-Me), 15.9, 16.6 (2-Me, 12-Me), 18.6 (5''-Me), 19.8, 19.9 (8-Me, 6-Me), 21.3, 21.5, 21.5 (C14, 6-Me, 3''-Me), 28.9 (C4'), 34.2 (C10), 34.9 (C2''), 36.1 (C8), 37.4 (C7), 39.1 (C4), 40.2 (NMe$_2$), 45.2 (C2), 49.4 (3''-OMe), 50.2 (6-OMe), 65.4 (C3'), 65.6 (C5''), 68.6 (C5'), 70.5 (C11), 71.1 (C2'), 72.7 (C3''), 74.8 (C12), 76.6 (C13), 78.0 (C4''), 78.6 (C6), 78.9 (C3), 80.5 (C5), 96.1 (C1''), 102.8 (C1'), 167.1 (C9), 175.9 (C1).

[33] McGill JM, Johnson R. Structural and conformational analysis of (*E*)-erythromycin A oxime. Magn Res Chem 1993, 31, 273–277.

[34] Xu Z-Q, Qiu Y-L, Chokekijchai S, Mitsuya H, Zemlicka J. Unsaturated acyclic analogs of 2'-deoxyadenosine and thymidine containing fluorine: synthesis and biological activity. J Med Chem 1995, 38, 875–882.

[35] The assays were performed in accordance with the Code of Federal Regulations guidelines: CFR Title 21, Part 436105.

[36] Schlunzen F, Zarivach R, Harms J, Bashan A, Tocilj A, Albrecht R, et al. Structural basis for the interaction of antibiotics with the peptidyl transferase centre in eubacteria. Nature 2001, 413, 814–821.

[37] Tuberculosis Antimicrobial Acquisition and Coordinating Facility (TAACF, Alabama, USA), cf. http://www.taacf.org.

[38] Thibessard A, Haas D, Gerbaud C, Aigle B, Lautru S, Pernodet J-L, et al. Complete genome sequence of *Streptomyces ambofaciens* ATCC 23877, the spiramycin producer. J Biotechnol 2015, 214, 117–118.

[39] Przybylski P. Modifications and biological activity of natural and semisynthetic 16-membered macrolide antibiotics. Curr Org Chem 2011, 15, 328–374.

[40] Liu L, Roets E, Hoogmartens J. Liquid chromatography of spiramycin on poly(styrene-divinylbenzene). J Chromatogr A 1997, 764, 43–53.

[41] Ramu K, Shringarpure S, Williamson JS. A solution conformation analysis of forocidins I and isoforocidins I using NMR and molecular modeling. Pharm Res 1995, 12, 621–629.

[42] Miura T, Kanemoto K, Natsume S, Atsumi K, Fushimi H, Yoshida T, et al. Novel azalides derived from 16-membered macrolides. Part II: Isolation of the linear 9-formylcarboxylic acid and its

sequential macrocyclization with an amino alcohol or an azidoamine. Bioorg Med Chem 2008, 16, 10129–10156.

[43] Hardy DJ, Hensey DM, Beyer JM, Vojtko C, McDonald EJ, Fernandes PB. Comparative *in vitro* activities of new 14-, 15-, and 16-membered macrolides. Antimicrob Agents Chemother 1988, 32, 1710–1719.

[44] Rubinstein E, Keller N. Spiramycin renaissance. J Antimicrob Chemother 1998, 42, 572–576.

[45] Pilot MA, Qin XY. Macrolides and gastrointestinal motility. J Antimicrob Chemother 1988, 22 (Suppl B), 201–206.

[46] Klich K, Pyta K, Kubicka MM, Ruszkowski P, Celewicz L, Gajecka M, et al. Synthesis, antibacterial, and anticancer evaluation of novel spiramycin-like conjugates containing C(5) triazole arm. J Med Chem 2016, 59, 7963–7973.

[47] Tang X, Dai H, Zhu Y, Tian Y, Zhang R, Mei R, et al. Maytansine-loaded star-shaped folate-core PLA-TPGS nanoparticles enhancing anticancer activity. Am J Transl Res 2014, 6, 528–537.

[48] Paterson I, Britton R, Delgado O, Gardner NM, Meyer A, Naylor GJ, et al. Total synthesis of (–)-dictyostatin, a microtubule-stabilising anticancer macrolide of marine sponge origin. Tetrahedron 2010, 66, 6534–6545.

[49] Shin Y, Fournier J-H, Brückner A, Madiraju C, Balachandran R, Raccor BS, et al. Synthesis and biological evaluation of (–)-dictyostatin and stereoisomers. Tetrahedron 2007, 63, 8537–8562.

[50] Wang L, Kitaichi K, Hui CS, Takagi K, Takagi K, Sakai M, et al. Reversal of anticancer drug resistance by macrolide antibiotics *in vitro* and *in vivo*. Clin Exp Pharmacol Physiol 2000, 27, 587–593.

[51] Bogdanov AA, Sumbatyan NV, Shishkina AV, Karpenko VV, Korshunova GA. Ribosomal tunnel and translation regulation. Biochemistry 2010, 75, 1501–1516.

[52] Poulsen SM, Kofoed C, Vester B. Inhibition of the ribosomal peptidyl transferase reaction by the mycarose moiety of the antibiotics carbomycin, spiramycin and tylosin[1]. J Mol Biol 2000, 304, 471–481.

[53] Hansen JL, Ippolito JA, Ban N, Nissen P, Moore PB, Steitz TA. The structures of four macrolide antibiotics bound to the large ribosomal subunit. Mol Cell 2002, 10, 117–128.

[54] Garza-Ramos G, Xiong L, Zhong P, Mankin A. Binding site of macrolide antibiotics on the ribosome: new resistance mutation identifies a specific interaction of ketolides with rRNA. J Bacteriol 2001, 183, 6898–6907.

[55] Depardieu F, Courvalin P. Mutation in 23S rRNA responsible for resistance to 16-membered macrolides and streptogramins in *Streptococcus pneumoniae*. Antimicrob Agents Chemother 2001, 45, 319–323.

[56] Lai CJ, Weisblum B. Altered methylation of ribosomal RNA in an erythromycin-resistant strain of *Staphylococcus aureus*. Proc Natl Acad Sci USA 1971, 68, 856–860.

[57] Mutak S, Maršić N, Kramarić MD, Pavlović D. Semisynthetic macrolide antibacterials derived from tylosin. Synthesis and structure–activity relationships of novel desmycosin analogues. J Med Chem 2004, 47, 411–431.

[58] Gebhardt P, Gräfe U, Möllmann U, Hertweck C. Semisynthetic preparation of leucomycin derivatives: introduction of aromatic side chains by reductive amination. Mol Diversity 2005, 9, 27–32.

[59] Przybylski P, Pyta K, Brzezinski B. Unexpected α,β-unsaturated products of reductive amination of the macrolide antibiotic josamycin. Tetrahedron Lett 2009, 50, 6203–6207.

[60] Zhao Z, Jin L, Xu Y, Zhu D, Liu Y, Liu C, Lei P. Synthesis and antibacterial activity of a series of novel 9-*O*-acetyl-4′-substituted 16-membered macrolides derived from josamycin. Bioorg Med Chem Lett 2014, 24, 480–484.

[61] Furuuchi T, Kurihara K, Yoshida T, Ajito K. Synthesis and biological evaluation of novel leucomycin analogues modified at the C-3 position I. Epimerization and methylation of the 3-hydroxyl group. J Antibiot 2003, 56, 399–414.

[62] Lazarova TI, Binet SM, Vo NH, Chen JS, Phan LT, Or YS. Synthesis of new 14-membered macrolide antibiotics *via* a novel ring contraction metathesis. Org Lett 2003, 5, 443–445.

[63] Li F, Yang B, Miller MJ, Zajicek J, Noll BC, Möllmann U, et al. Iminonitroso Diels−Alder reactions for efficient derivatization and functionalization of complex diene-containing natural products. Org Lett 2007, 9, 2923–2926.

[64] Yang B, Zöllner T, Gebhardt P, Möllmann U, Miller MJ. Preparation and biological evaluation of novel leucomycin analogs derived from nitroso Diels−Alder reactions. Org Biomol Chem 2010, 8, 691–697.

[65] Carosso S, Miller MJ. Nitroso Diels−Alder (NDA) reaction as an efficient tool for the functionalization of diene-containing natural products. Org Biomol Chem 2014, 12, 7445–7468.

[66] Sano H, Tanaka H, Yamashita K, Okachi R, Omura S. Chemical modification of spiramycins V. Synthesis and antibacterial activity of 3′- or 4‴-de-*N*-methylspiramycin I and their *N*-substituted derivatives. J Antibiot 1985, 38, 186–196.

[67] Adamski RJ, Heymann H, Geftic SG, Barkulis SS. Preparation and antibacterial activity of some spiramycin derivatives. J Med Chem 1966, 9, 932–934.

[68] Seki A, Mori T, Sasaki K, Takahashi Y, Miyake T, Akamatsu Y. Synthesis of 17-membered azalides from a 16-membered macrolide utilizing amide-selective silane reduction. Med Chem Comm 2015, 6, 581–585.

[69] Miura T, Natsume S, Kanemoto K, Atsumi K, Fushimi H, Sasai H, et al. Novel azalides derived from sixteen-membered macrolides. J Antibiot 2007, 60, 407–435.

[70] Miura T, Natsume S, Kanemoto K, Shitara E, Fushimi H, Yoshida T, et al. Novel azalides derived from 16-membered macrolides. III. Azalides modified at the C-15 and 4″ positions: improved antibacterial activities. Bioorg Med Chem 2010, 18, 2735–2747.

[71] Seiple IB, Zhang Z, Jakubec P, Langlois-Mercier A, Wright PM, Hog DT, et al. A platform for the discovery of new macrolide antibiotics. Nature 2016, 533, 338–345.

[72] Testa C, Scrima M, Grimaldi M, D'Ursi AM, Dirain ML, Lubin-Germain N, et al. 1,4-Disubstituted-[1,2,3]triazolyl-containing analogues of MT-II: design, synthesis, conformational analysis, and biological activity. J Med Chem 2014, 57, 9424–9434.

[73] Caselli E, Romagnoli C, Vahabi R, Taracila MA, Bonomo RA, Prati F. Click chemistry in lead optimization of boronic acids as β-lactamase inhibitors. J Med Chem 2015, 58, 5445–5458.

[74] Ran X, Liu L, Yang C-Y, Lu J, Chen Y, Lei M, et al. Design of high-affinity stapled peptides to target the repressor activator protein 1 (RAP1)/telomeric repeat-binding factor 2 (TRF2) protein–protein interaction in the Shelterin complex. J Med Chem 2016, 59, 328–334.

[75] Hirose T, Sunazuka T, Noguchi Y, Yamaguchi Y, Hanaki H, Barry Sharpless K, et al. Rapid "SAR" *via* click chemistry: an alkyne-bearing spiramycin is fused with diverse azides to yield new triazole-antibacterial candidates. Heterocycles 2006, 69, 55–61.

[76] Klich K, Pyta K, Przybylski P. Regio- and stereoselective functionalization of 16-membered lactone aglycone of spiramycin *via* cascade strategy. J Org Chem 2015, 80, 7040–7049.

[77] Kannan K, Mankin AS. Macrolide antibiotics in the ribosome exit tunnel: species-specific binding and action. Ann NY Acad Sci 2011, 1241, 33–47.

Index

ABT-773 83
12-acylimidazolyl clarithromycin 47
8,9-anhydroerythromycin A 6,9-hemiketal 33
9a-aza-9-deoxo-9,9-dihydro-9a-
    homoerythronolide A 22
9a-aza-9-deoxo-9a-homoerythromycin A 82
9a-sulfonyloxime 40
acetic acid 53
acylation 89
ADMET 64
aglycone 19, 21, 89, 90
alkyl halide 37, 40
alkylation 35, 36, 40, 49, 50, 57, 83
alkyne 89
allyl bromide 50
allylation 50, 51
allyloxylation 57
anhydroerythromycin A 2–5, 23, 33
antibacterial activity 15, 37, 42, 44, 45, 50, 53,
    57, 87, 92
antibacterial macrolide hybrid 56
anticancer activity 92, 93
aprotic solvents 23
association 16
azahomoerythromycin 70
azalide 38, 57
azithromycin 1–3, 6, 7, 9, 13–15, 19–21, 31, 32,
    38, 40–45, 66–72, 82, 83, 87
azithromycin hybrid 42

Bacillus subtilis ATCC6633 83
bacterial membrane 17
bacterial ribosomes 10, 12, 16, 17, 44, 64
– translation 64
– genetic code 64
Beckmann rearrangement 38, 40, 44
benzylchloroformate 36
bile acid 17, 70–72
bile micelles 70, 72
binding epitopes 67
bioactivity 70
bioavailability 70
biosynthesis 1, 23
bound conformation 65, 67, 69

calorimetry 63
Campylobacter coli 72
Campylobacter jejuni 72
capillary electrophoresis 63
carbomycin 9
catalytic hydrogenation 36, 40, 41, 86
catalytic hydrogenolysis 40
catalytic reduction 40
cell membranes 70
cellular permeability 18
cethromycin 32, 49–53
chemical properties 23
cholestasis 71
cladinose 15, 19, 20, 32, 45, 70
clarithromycin 1–3, 5–7, 9, 11, 19, 31, 32, 35–39,
    42, 44, 45, 47, 53, 69, 70, 72, 83, 84, 87
clarithromycin 10,11-cyclic carbamate 46
clarithromycin 11,12-cyclic carbamate 47
clarithromycin 2',4''-diacetate 86
clarithromycin 9-E-oxime 44
clarithromycin B 9
clarithromycin-adenine 83
cleavage conjugates 79
conformational analysis 23
conformational flexibility 23
conformational state 19, 65
– folded-out 19–23, 69
– folded-in 19–23
conjugate 32, 79
constrained molecular mechanics 23
Corey-Kim method 53
Corey-Kim oxidation 56
Corey-Kim reagent 53
coupling constant 19, 22, 23
crystal structure 17, 22, 23, 64, 65, 68, 69
crystallography 63
cyclic oximes 38
cycloaddition 53, 58
cytotoxic activity 92

(-)-dictyostatin 88
1,8-diazabicyclo[5.4.0]undec-7-ene 47
11-O-dimethyl-9a-homoerythronolide A 22
6-deoxyerythronolide B 33

https://doi.org/10.1515/9783110515756-005

8-*deuterio*-erythromycin B 9
9-deoxo-8a-aza-8a-homoerythromycin
    A 42, 43
9-deoxo-9a-aza-9a-homoerythromycin 40, 41
9-deoxo-9a-methyl-9a-aza-9a-
    homoerythromycin A 40
decladinosylazithromycin 70
dehydration 7
Deinococcus radiodurans 9–13, 68
demethylation 80
deoximation 50
deoxycholate micelles 72
desosamine 12, 15, 19, 20, 23, 32, 68, 70,
    81, 82
Dess-Martin oxidation 53
diazotation 57
Diels-Alder reaction 89
diffusion ordered spectroscopy (DOSY) 7, 70
dirithromycin 19
dissociation 16
dissolution 18
dodecylphosphocholine 70
domain II 89
domain V 64
drug design 17, 18, 63, 73, 79
drug discovery 63, 79

(*E*)-9-hydroxyimino-6-*O*-methylerythronolide
    A 21
1-ethyl-3-(3-dimethylaminopropyl)carbodiimide
    hydrochloride 47
3-*endo* folded-out 21
E. coli 11, 16, 18, 22, 66, 67, 69, 80, 89
efflux 64
epoxidation 89
equilibrium dialysis 16, 63
erythromycin A 1–7, 9, 10–13, 15, 16, 19, 22, 23,
    31–37, 40, 42, 45, 50, 57, 58, 64, 67, 69–71,
    80, 81, 83
– acid treatment 3, 5
– equilibrium mixture 3
erythromycin A 2'-ethyl succinate 5, 8
erythromycin A 9,11-iminoether 40
erythromycin A 9-*E*-oxime 36, 38, 40, 42, 44,
    53, 54
erythromycin A hydroiodide 19
erythromycin A-6,9-iminoether 40, 41
erythromycin B 2, 7, 8, 33
erythromycin B enol ether ethyl succinate 7–9
erythromycin B ethyl succinate 7, 9

erythromycin C 2, 33
erythromycin class antibiotics 10
erythromycin D 2, 33
erythromycin E 2
erythromycin enol ether 2, 3
erythromycin F 2
erythromycin G 2
Eschweiler-Clarke condition 42
Eschweiler-Clarke methylation 44
etherification 89
ethylenediamine 53
exit tunnel 12, 13, 18, 64, 68, 69, 80, 89

fermentation process 33
filtration 17
fluorescence spectroscopy 63, 71
fluorination 53
Fokin-Huisgen cycloadditions 89
force field 23
forosamine 88, 89
FT-IR 24, 90
fused hybrid molecules 79

glacial acetic acid 40
global energy minimum 24
global minimum conformation 23
glucosidase DesR 72
Gram-positive bacteria 92, 93

Haemophilus influenzae 57
half-life 4, 5
Haloarcula marismortui 9, 10, 13, 22, 23, 68,
    89, 92
Heck coupling 50
HMBC 90
HSQC 90
hybride macrolide 59
hydrogen bonding interactions 12
hydrogenation 57
hydrolysis 7, 8

inactivation 64
intestinal absorption 18
intramolecular ketalization 35
intramolecular Michael addition 50

josamycin 23, 31

ketolide 45, 57, 67
kinetic data 3

L4 ribosomal protein 13
lactone ring 15, 21, 23
lead tetraacetate 59
lecithin liposomes 72
leucomycin 89
ligand 66
ligand-receptor complexes 63
ligand-receptor interactions 63, 70
lincosamides (L) 80
lipid membranes 72
Lipinski's Rule-of-Five 32
lysosomal membranes 70

16-membered macrolides 23
6-O-methyl azithromycin 45
6-O-methyl homoerythromycin 21
6-O-methyl roxithromycin 45
macrocyclization 58
macrolide hybrid compounds 80
macrolide ring 19
macrolone 57, 59
maytansine 88
merged hybrid 32, 79
metal hydrides 41
metathesis 89
methanolysis 53, 56
methyl iodide 35, 36, 42
methyl transferase enzymes 80
methyl transferases 64
methylation 36, 41, 42, 80
methymycin 72
Micrococcus luteus ATCC 9341 87
Mitsunobu reaction 82
modithromycin 32, 53, 54, 56
molecular dynamics simulations 22
molecular mechanics calculations 24
molecular modeling 19–23, 69
monomethylation 80
mutations 64
Mycobacterium tuberculosis H37Rv 83, 88
mycoplasmas 88

3'-N-bis(benzyloxycarbonyl)-N-demethyl
    azithromycin 42, 43
3'-N-bis(benzyloxycarbonyl)-N-demethyl
    erythromycin A 35
3'-N-demethyl-6-O-methylerythromycin A 36
N-alkylation 59
neutral macrolide 16
N-fluorobenzenesulfonimide 53

N-methylation 89
NMR 3–5, 7, 19–24, 63, 65, 67, 69, 71, 86, 90
NMR chemical-shift titration experiments 72
NMR translational diffusion 70
NOE 19–21, 23, 24, 65
NOE constraints 24
NOESY 21, 67, 90
nucleic acids 63
nucleobases 81, 82
nucleophilic addition 89
nucleosides 81

6-O-allyl erythromycin A 50
6-O-methyl azithromycin 44
oleandomycin 20, 22, 31, 66, 70
oxidation 53
oximation 42

2-pyrimidine 38
4-(3-pyridinyl)-1H-imidazole-1-butanamine 47
paramagnetic relaxation enhancements
    (PREs) 70, 72
peptidyl transferase activity 10, 17
peptidyl transferase cavity 11, 12
peptidyl transferase region 64
peptidyl-transferase tunnel 87
peptidyl-tRNAs 10, 12
permeability 70
Pfitzner-Moffat procedure 47
pharmacophore 67, 79
phosphatidylmembrane 72
phospholipase A1 70–72
phospholipidosis 70, 71
phospholipids 70
plasma 7
plasma proteins 70
platinum oxide 40
PM5 method 24
potassium *tert*-butoxide 37
pro-drug 8
propenyl quinolyl *t*-butylcarbonate 50
protein L4 64
protein synthesis 80
proteins 63, 64
protonated macrolide 16

radiolabelled antibiotics 17
receptor 66, 70
reduction 89
reductive amination 59

reductive coupling 58
reductive methylation 36
reprotection 50
resistance 64, 73, 80, 89
resistant strains 80
ribosomal proteins 13
ribosome 16, 21, 64, 65, 67, 69, 80
ribosome binding 15
– rate constants 15
– binding affinity 15
– rate of uptake 15
– macrolide-ribosome complexes 15, 17, 64, 67
– kinetics 15
– weak interaction 16
– strong interaction 16
– magnesium 16
ribosome binding modes 18
ring folding 21
RNA footprinting 17, 63
ROE 19
ROESY 21–24
roxithromycin 1, 2, 5, 7, 9, 11, 19, 31, 32, 34, 35, 67, 71, 83, 84
RU 72366 22, 23

(9S)-erythromycylamine A 81
23S rRNA 11, 13, 14, 64, 80, 89
Saccharopolyspora erythraea 1, 31, 33
saturation 66
secondary amine 59
self-diffusion measurements 72
semi empirical PM3 calculations 24
semisynthesis 36
silylation 36
small-angle X-ray scattering (SAXS) 72
sodium bis(trimethylsilyl)amide 47
sodium borohydride 41
sodium dodecylsulphate (SDS) micelles 70
sodium hydride 35
solithromycin 32, 53, 54
solubility 70
solution-state structure 17, 18
solvent paramagnetic relaxation
    enhancement 70
Sonogashira method 57
Sorangium cellulosum 24
spin-diffusion mechanism 66

spiramycin 1, 2, 9, 31, 88–90, 92, 93
spiramycin I 88
spiramycin II 88
spiramycin III 88
spiramycins I–III 88
STD 65–67, 69, 70, 72
sterical blocking 10
stomach acid 6
stomach fluid 33
Streptococcus pneumonia 44
Streptococcus pyogenes 57
streptogramin (S$_B$) 80
Streptomyces ambofaciens 88
Streptomyces spp. 31
structure 23
sugar unit 19
sulphonylation 89
superimposition 23
surface plasmon resonance 17

taurocholic acid 72
TE-802 32, 53, 55
telithromycin 1, 2, 7, 9, 22, 23, 32, 47–50, 53, 61, 65, 67, 70, 83, 84
Thermus thermophilus 13
thuggacin A 24
thuggacin C 24
thymidine 83
thymidine derivatives 82
tianchimycin A 24
tosyl chloride 40
Toxoplasma gondii 88
translactonisation 81
tritylhydrazone 38
trNOE 65
trNOESY 65, 69
tylosin 9, 23, 69, 89

ultrafiltration 63
unconstrained molecular modeling 23

vicinal coupling constant 21, 24

ximelagatran 72
X-ray crystallography 16
X-ray diffraction 19, 63